Regional Ground-Water Evapotranspiration and Ground-Water Budgets, Great Basin, Nevada

By William D. Nichols

U.S. GEOLOGICAL SURVEY PROFESSIONAL PAPER 1628

Prepared in cooperation with the
LAS VEGAS VALLEY WATER DISTRICT and the
NEVADA DIVISION OF WATER RESOURCES

AF854963

Cover photograph: Southern Snake Range, White Pine County, Nevada, August 1997. View east across southern Spring Valley from U.S. Route 93. Lincoln Canyon Creek meanders across alluvial fan in center middle ground. Massive cliff-forming rocks of Pole Canyon Limestone formation (Middle Cambrian age) cap Mount Washington, on far left of view, and ridge that descends southward (to right). Rocks of Upper Cambrian Notch Peak Limestone formation cap ridges on skyline at center of view. These limestones (or their stratigraphic equivalents) form part of lower carbonate aquifer in subsurface. Ground-water evapotranspiration from phreatophytes and associated bare soil in 1,667-square-mile Spring Valley may total about 90,000 acre-feet per year. Photograph by W.D. Nichols, U.S. Geological Survey.

U.S. DEPARTMENT OF THE INTERIOR
BRUCE BABBITT, Secretary

U.S. GEOLOGICAL SURVEY
CHARLES G. GROAT, *Director*

The use of firm, trade, and brand names in this report is for identification purposes only and does not constitute endorsement by the U.S. Government.

Reston, Virginia 2000

Library of Congress Cataloging in Publication Data

Nichols, William D.
Regional ground-water evapotranspiration and ground-water budgets, Great Basin, Nevada / by William D. Nichols; prepared in cooperation with the Las Vegas Valley Water District and the Nevada Division of Water Resources.
p. cm. -- (U.S. Geological Survey professional paper ; 1628)
Includes bibliographical references.
ISBN 0-607-94492-7
1. Groundwater flow--Nevada. 2. Evapotranspiration--Nevada 3. Phreatophytes--Nevada. I. Nevada. Division of Water Resources II. Las Vegas Valley Water District (Nev.) III. Title. IV. Series.

GB1197.7 .N53 2000
551.49'09793--dc21

00-034081

For sale by the U.S. Geological Survey, Information Services, Box 25286
Denver Federal Center, Denver, CO 80225

FOREWORD

Ground water is the main source of water supply in much of Nevada and the Great Basin. The ground-water budget is fundamental to quantitative analyses of this resource. In the topographically closed valleys of the region, ground-water recharge is from precipitation and interbasin ground-water flow from adjacent valleys. The principal mechanisms of ground-water discharge are the evapotranspiration of ground water by phreatophytes and interbasin ground-water flow to adjacent valleys. Interbasin ground-water flow between adjacent valleys underscores the lack of coincidence of topographic and hydrologic basins throughout the study area, which is underlain by thick sequences of carbonate rocks.

Previous Nevada ground-water budgets used results of field studies in California and Utah published in 1912 and 1932 to estimate ground-water evapotranspiration. Assuming steady-state conditions, with no interbasin ground-water flow to or from a topographic basin, ground-water recharge by precipitation was assumed to be equal to the estimated ground-water evapotranspiration. If the ground-water budget for a given valley proved to be imbalanced, then the difference was reconciled by assigning interbasin flow to or from the valley if an appropriate opposite imbalance existed in an adjacent valley. If interbasin ground-water flow could not be invoked, then the budget was either modified to achieve balance or was not balanced.

The principal problem with the application of this method of estimating ground-water budgets lay in the studies upon which the estimates of ground-water evapotranspiration were based. The 1912 and 1932 studies were outstanding pieces of scientific work, but the results were overextended when they were used to make regional-scale estimates of ground-water evapotranspiration. New studies of ground-water evapotranspiration in phreatophytic zones using micrometeorological methods at remote field sites were begun by the U.S. Geological Survey in Nevada in 1988. The results of these studies, together with results of similar studies in Owens Valley, Calif., were used to develop the relation between ground-water evapotranspiration and plant cover, which is described in Chapter A of this report. Using the functional form of this relation, an estimate of ground-water evapotranspiration can be made from an estimate of plant cover.

Large areas in the Great Basin have phreatophytic zones that have a plant cover of 20 percent or less. Evapotranspiration rates from these areas are low, but the volume of ground water evapotranspired on a regional scale is large because of the large areas involved. Plant cover can be determined readily, on a regional scale, from Landsat data using easily calculated vegetation indices. The most commonly used vegetation index, the normalized-difference vegetation index, was not sufficiently sensitive to the sparse vegetation conditions that characterize the Great Basin. Field measurements of plant cover were used to develop an improved relation between plant cover and a modified soil-adjusted vegetation index, derived from satellite data, that is more sensitive to sparse vegetation. This improved functional relation between the vegetation index and plant cover was used together with the relation between plant cover and ground-water evapotranspiration described in Chapter A to estimate ground-water evapotranspiration at a regional scale as discussed in Chapter B.

These tools provided the means for estimating ground-water evapotranspiration at regional scales. A study area covering nearly 15,000 mi^2 and including 16 contiguous valleys of eastern Nevada in the central Great Basin was selected to apply the methods described in Chapters A and B. Estimates of ground-water evapotranspiration were determined for each valley for 1985 and 1989, and a mean annual estimate was calculated. Ground-water recharge then was estimated by using ground-water evapotranspiration estimates from 15 of the valleys to determine recharge coefficients to be applied to mean annual precipitation over the entire region. These efforts and the results are described in Chapter C. The estimates of ground-water recharge and discharge were used to develop ground-water budgets for each valley. As with previous studies, any imbalance in a given valley was assumed to be corrected by interbasin flow between adjacent valleys with opposite imbalances.

Estimates of interbasin ground-water flow require reasonably good knowledge of ground-water levels, local geology, and the hydraulic properties of the local unconsolidated and consolidated aquifers, but these data generally are sparse throughout Nevada. The paucity of these data was nearly as great for this study as it was for previous studies made two or more decades ago. However, the present study benefited from the hydrologic knowledge developed by the earlier studies and from geologic mapping that has occurred in the area during the past few decades. Consequently, as described in Chapter C, interbasin flow to balance the ground-water budget for a given valley is proposed only in areas where it was suggested by earlier studies or in areas where it is supported by available geologic and hydrologic data.

CONTENTS

COMPACT DISC (in pocket at back of report)

Contains electronic versions of Professional Paper 1628 (this report) and two related publications:

- U.S. Geological Survey Open-File Report 99-242, Data Sets and Related Information Used for Estimating Regional Ground-Water Evapotranspiration in Eastern Nevada, by J. LaRue Smith, Brian D. Reece, and Rose L. Medina
- U.S. Geological Survey Fact Sheet 073-00, New Regional Ground-Water Budgets for Nevada, by William D. Nichols and A.S. VanDenburgh

CONVERSION FACTORS

Inch-pound units of measure used in this report may be converted to International System of units (SI) by using the following factors

Multiply	By	To obtain
Area		
acre	4,047	square meter
square foot (ft^2)	0.09290	square meter
Length		
foot (ft)	0.3048	meter
inch (in.)	2.540	centimeter
mile (mi)	1.609	kilometer
square mile (mi^2)	2.590	square kilometer
Volume		
acre-foot (acre-ft)	1,233	cubic meter
acre-foot per year (acre-ft/yr)	1,233	cubic meter per year
Flow rate		
foot per day (ft/d)	0.3048	meter per day
foot per year (ft/yr)	0.3048	meter per year
foot squared per day (ft^2/d)	0.09290	meter squared per day

Temperature: Degrees Fahrenheit (°F) may be converted to degrees Celsius (°C) by using the formula °C = [°F - 32]/1.8.

Sea level: In this report, "sea level" refers to the National Geodetic Vertical Datum of 1929 (NGVD of 1929, formerly called "Sea-Level Datum of 1929"), which is derived from a general adjustment of the first-order leveling networks of the United States and Canada.

Chapter A. Determining Ground-Water Evapotranspiration From Phreatophyte Shrubs and Grasses as a Function of Plant Cover or Depth to Ground Water, Great Basin, Nevada and Eastern California

By William D. Nichols

Title-page photograph: Southern Ruby Valley, Elko County, Nevada, July 1996. View west toward southern Ruby Mountains. Sharp color contrast marks boundary between phreatophyte greasewood (bright green) and non-phreatophyte sagebrush (dark green and foreground). Ruby Marsh is in middle distance at base of mountains. Phreatophytes and associated bare soil on floor of Ruby Valley may yield almost 170,000 acre-feet of ground-water evapotranspiration per year. Photograph by William D. Nichols, U.S. Geological Survey.

CONTENTS

FIGURES

TABLES

Chapter A. Determining Ground-Water Evapotranspiration From Phreatophyte Shrubs and Grasses as a Function of Plant Cover or Depth to Ground Water, Great Basin, Nevada and Eastern California

By William D. Nichols

ABSTRACT

Ground-water evapotranspiration data from five sites in Nevada and seven sites in Owens Valley, California, were used to develop equations for estimating ground-water evapotranspiration as a function of phreatophyte plant cover or as a function of the depth to ground water. Equations are given for estimating mean daily seasonal and annual ground-water evapotranspiration. The equations that estimate ground-water evapotranspiration as a function of plant cover can be used to estimate regional-scale ground-water evapotranspiration using vegetation indices derived from satellite data for areas where the depth to ground water is poorly known. Equations that estimate ground-water evapotranspiration as a function of the depth to ground water can be used where the depth to ground water is known, but for which information on plant cover is lacking.

INTRODUCTION

Ground-water evapotranspiration by phreatophytes and associated bare soil is an important component of the water budget in arid and semiarid regions. In many valleys of the Great Basin, ground-water evapotranspiration is the only mechanism of ground-water discharge or represents the major component of discharge in the ground-water budget. Estimating ground-water evapotranspiration, especially at regional scales, has been difficult at best.

Estimates of ground-water evapotranspiration most commonly have been made either for specific plant types using evapotranspiration-tank data (Lee, 1912; Blaney and others, 1930, 1938; White, 1932; Young and Blaney, 1942; Gatewood and others, 1950; Robinson, 1970), or for small areas of native rangeland or riparian vegetation based on field measurements using micrometeorological techniques (Weeks and others, 1987; Duell, 1990; Malek and others, 1990). These estimates were based on measured evapotranspiration derived from energy budgets and were then extrapolated to larger areas of similar vegetation.

The phreatophytes of interest are those of the salt desert plant community, which include saltgrass (*Distichlis spicata* var. *stricata*), iodinebush (*Allenrolfea occidentalis*, also called pickleweed), and saltsage (*Atriplex tridentata*); and those of the shadscale-greasewood plant community, which include greasewood (*Sarcobatus vermiculatus*), shadscale (*Atriplex confertifolia*), saltbush (*Atriplex canescens*, also known as chamiso, and *Atriplex torreyi*), spiny hopsage (*Grayia spinosa*), winterfat (*Ceratoides lanata*), and, where soils and ground water are less saline, rabbitbrush (*Chrysothamnus nauseosus*) and big sagebrush (*Artemisia tridentata*).

Saltgrass, the principal phreatophyte of the salt desert community, commonly grows where the depth to ground water is less than about 8 ft, but has been reported to grow in areas where the water table is as much as 12 ft deep (Blaney and others, 1933, p. 50). Plants of the shadscale-greasewood community occur where depths to water range from about 3 ft to as much as 65 ft (Robinson, 1958). Greasewood, the principal phreatophyte of this plant community, occurs in areas where the depth to ground water ranges from about 5 ft to 35 ft and perhaps as much as 60 ft (Robinson, 1958, p. 39). Saltbush is found where the water table is from about 8 ft to as much as 62 ft below land surface (Robinson, 1958, p. 33). Rabbitbrush grows where the depth to water is less than about 35 ft (Robinson, 1958, p. 34).

Big sagebrush, although commonly not considered to be a phreatophyte, has been observed growing in many areas in association with rabbitbrush where the water table is as much as 12 ft below land surface (Mozingo, 1987, p. 271). Under these conditions of shallow ground water, big sagebrush is assumed to be a phreatophyte.

Purpose and Scope

The present study extends earlier work on ground-water evapotranspiration by phreatophytes (Nichols, 1994) and this chapter presents equations for estimating ground-water evapotranspiration from phreatophyte grasses and shrubs and from associated bare soil in areas of shallow ground water as a function of plant cover or depth to ground water. The equations previously proposed (Nichols, 1994) required knowledge of the amount of plant cover as well as the depth to ground water. Depth to ground water commonly is not well known at regional scales in many areas, such as the Great Basin, and is difficult to define in the absence of a sufficient number of wells. Equations proposed by the present study that correlate ground-water evapotranspiration with plant cover can be used to estimate ground-water evapotranspiration at regional scales using remotely sensed vegetation index data, provided correlation can be made between plant cover and satellite-data-derived vegetation indices. The equations that describe ground-water evapotranspiration as a function of depth to ground water are appropriate for use, in combination with the equations given by Nichols (1994), in numerical models of ground-water flow and for other applications where the depth to ground water is known.

Previous Studies

All evapotranspiration-tank studies of ground-water evapotranspiration used similar techniques. Tanks for some of the studies were placed in excavations and filled with the excavated soil; the soil surface in the tank either received transplanted greasewood, rabbitbrush, or other shrubs or was covered with saltgrass sod (Lee, 1912; White, 1932). For other studies, such as that at Santa Ana, Calif. (Blaney and others, 1930), the tank was driven into the ground so that it contained undisturbed soil and saltgrass. This method required an excavation around the tank as work proceeded. The filled tank then was lifted out of the excavation, the bottom of the tank was sealed, and the tank was placed in an excavation at the study site. The "tanks" employed by Robinson (1970) were not tanks as such, but rather were rectangular excavations measuring 30 ft by 30 ft by 10.5 ft deep, 20 ft by 20 ft by 10 ft deep, and 10 ft by 10 ft by 7 ft deep, that were lined with a heavy plastic membrane. In all the studies, the evapotranspiration tanks were fitted with one or two water-reservoir tanks. The reservoir tanks were used to maintain a constant water level in the evapotranspiration tank and to provide a storage tank to which water could be added in measured volumes. A record of volume added and volume remaining in the tanks provided a measure of the volume of evapotranspiration from the tank.

The pioneering study of evapotranspiration from saltgrass by Lee (1912) seemed to demonstrate a clear relation between ground-water evapotranspiration and depth to ground water. Subsequent studies by Blaney and others (1930, 1938) and studies summarized by Young and Blaney (1942) produced similar results, but analysis of these data (Weeks and others, 1987) demonstrated that depth to ground water does not adequately define a unique relation to ground-water evapotranspiration by saltgrass. None of these tank studies provided a measure of the amount of vegetation, such as density or leaf area index.

Evapotranspiration-tank experiments by White (1932) in the Escalante Desert near Milford, Utah, and Robinson (1970) in the Great Basin Desert near Winnemucca, Nev. (fig. A1), included evapotranspiration measurements for greasewood and rabbitbrush. These studies, as well as that of Gatewood and others (1950), attempted to relate measured evapotranspiration to the amount of vegetation present and to the depth to ground water. White (1932) developed a relation between the observed volume of ground-water evapotranspiration and the weight of dry biomass produced. White concluded a general relation existed between the depth to ground water and ground-water evapotranspiration by greasewood by assigning one rate of evapotranspiration to areas where ground water was less than 8 ft below land surface and a lower evapotranspiration rate to areas where ground water was greater than 8 ft deep. He also concluded that greasewood, rabbitbrush, and shadscale had similar rates of ground-water evapotranspiration. Gatewood and others (1950) and Robinson (1970) each developed measures of ground-water evapotranspiration as related to the volume density of vegetation, but each used different methods for determining volume density. The conclusions reached by Robinson (1970) are equivocal and not easily summarized, but suggest a relation between ground-water evapotranspiration and both the depth to ground water and the volume of biomass. None of these studies developed methods that can be applied systematically to estimate ground-water evapotranspiration by phreatophytes.

Energy budget studies, such as those by Gay and Fritschen (1979), Weeks and others (1987), Duell (1990), Malek and others (1990), and Nichols and others (1997), typically used either the Bowen ratio method (Tanner, 1960) or the eddy correlation method (Businger and others, 1967) to calculate evapotranspiration. Results of these studies apply to an area within about 300 to 500 ft of the point of measurement. Evapotranspiration measured by these studies commonly includes not only ground-water evapotranspiration, but water removed by evapotranspiration of any recent precipitation.

Ground-water transpiration by phreatophyte shrubs in the Great Basin was estimated by Nichols (1994) using an energy-combination model that solved the energy budget separately for the soil and the canopy. Nichols (1994) developed a functional relation between the transpiration of ground water, and the depth to ground water, plant density, and leaf area index. This approach did not include estimates of ground-water evaporation from the bare soil associated with the phreatophyte shrubs, and provided only for estimates of ground-water transpiration during the summer months, May through September. The results, as presented (Nichols, 1994), implied that leaf area index was independent of the depth to ground water.

DATA AND METHODS

Energy budget studies of ground-water evapotranspiration from native rangelands were done in Nevada from 1988 through 1994 (Nichols, 1994; Nichols and others, 1997). The field sites were in the Smoke Creek Desert, Smith Creek Valley, Railroad Valley (Nichols, 1994), and the Ash Meadows area (Nichols and others, 1997). Energy budget data for seven field sites in the Owens Valley of California (Duell, 1990) also were used in this study. Site locations are shown in figure A1 and site descriptions are given in tables A1 and A2. Methods of data collection and analysis for the Nevada sites have been discussed by Nichols (1994) and for the California sites by Duell (1990). Evapotranspiration for the Nevada sites was determined using the energy budget - Bowen ratio method (Tanner, 1960). For the California sites, Duell (1990, p. E25) determined evapotranspiration using either the Bowen ratio method or from eddy correlation data using a direct-measurement method, a residual method, or both; whenever two methods were used, the results were averaged for the present study.

The present analysis assumed, as have previous investigators, that ground-water evapotranspiration by phreatophytes is related to depth to ground water and plant biomass, but also assumed that biomass is a function of the depth to ground water. Factors other than the depth to ground water, such as soil type and soil and water chemistry, also may affect the biomass of any given area. However, in general, the shallower the depth to ground water, the more dense is the phreatophyte biomass. The amount of phreatophyte biomass, not the depth to ground water, is the principal indicator of ground-water evapotranspiration. If phreatophyte biomass is a function of the depth to ground water, and if ground-water evapotranspiration depends on phreatophyte biomass, then only the relation between phreatophyte biomass and ground-water evapotranspiration needs to be determined.

The measure of phreatophyte biomass used in this study is plant cover, which was determined from plant density and leaf area index. Plant density (*d*) was measured and is the ratio of the horizontal length of plants per length of measured transect, usually a length of 300 ft (Smith and others, 2000). Shrub leaf area index (*LAIp*) is the leaf area per unit ground area for individual shrubs along the measured plant-density transect and is a commonly used measure that typically ranges from 0 to 4, but has been reported as high as 10 for cereals with vertical leaf habits (Monteith and Unsworth, 1990). The *LAIp* used for this analysis was the measured or estimated annual maximum plant leaf area index at each field site ($(LAIp)_{max}$). For shrubs and grasses at the Nevada field sites, and generally throughout the Great Basin, the $(LAIp)_{max}$ is attained in mid-June to early July. The leaf area index (*LAI*) at the Nevada field sites was determined by measuring the assumed $(LAIp)_{max}$ of individual shrubs (Groeneveld and Warren, 1992) along a measured transect, calculating a weighted average of the $(LAIp)_{max}$ based on the percent of each shrub species along the transect $(\overline{LAIp})_{max}$, and multiplying $(\overline{LAIp})_{max}$ by the density of shrubs along the transect (eq. 1). For the Owens Valley sites, the *LAI* was estimated based on the plant

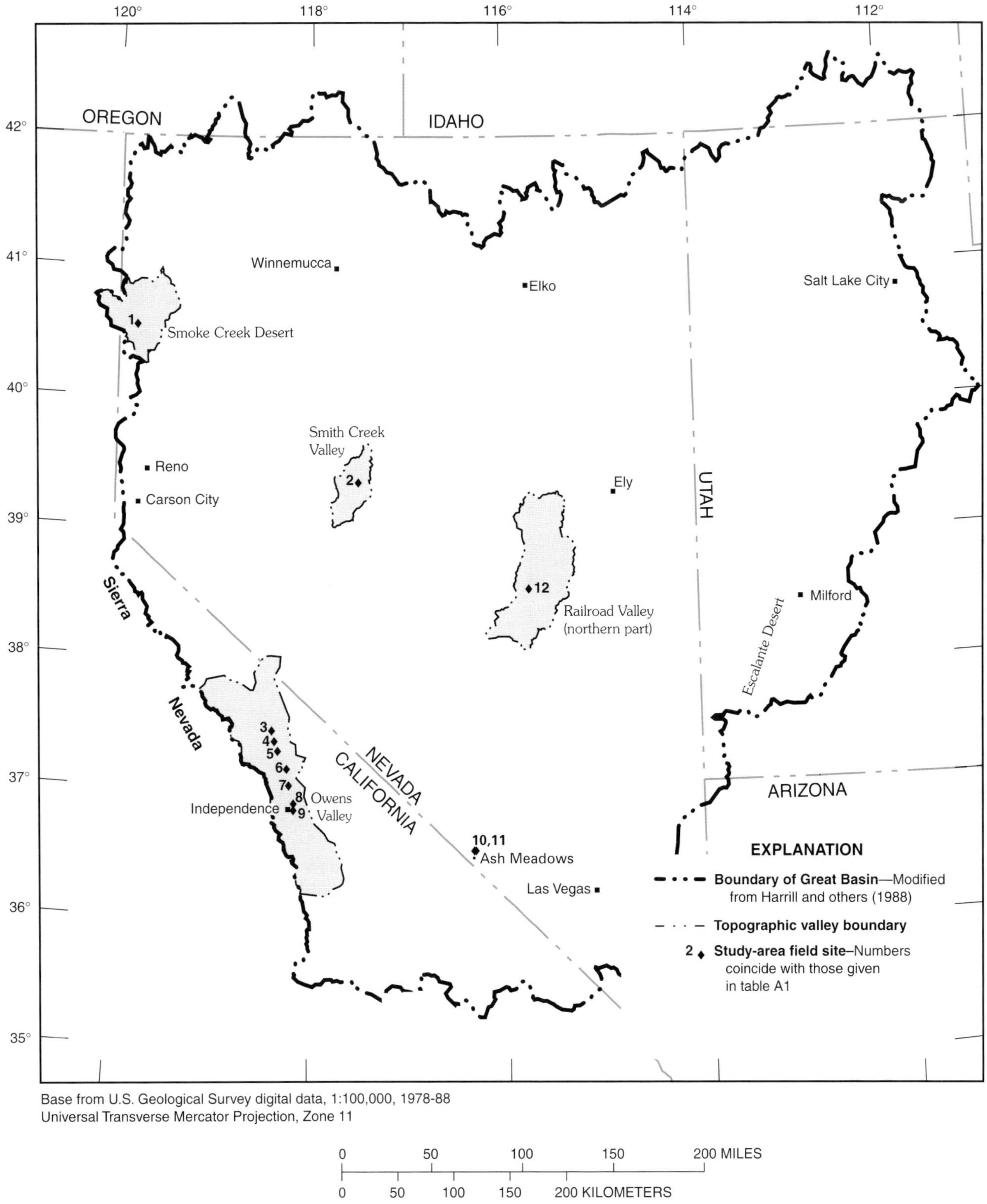

Figure A1. Location of study areas and field sites, Nevada and California. Numbers refer to locations given in table A1.

Table A1. General information for field sites, Great Basin study areas

[NR: Altitude not reported, but all sites are between 3,800 and 4,200 feet above sea level (Duell, 1990, p. E2).]

Site (fig. A1)	Location	Latitude (°N)	Longitude (°W)	Altitude (feet above sea level)	Dates of data collection	Source
1	Smoke Creek Desert, Nev.	40.534	119.818	3,907	June-Sept. 1991	Nichols, 1994
2	Smith Creek Valley, Nev.	39.330	117.512	6,046	May-Sept. 1989	Nichols, 1994
3	Owens Valley, Calif., site A	37.400	118.383	NR	Jan. 1984-Oct. 1985	Duell, 1990
4	Owens Valley, Calif., site C	37.317	118.367	NR	Jan. 1984-Oct. 1985	Duell, 1990
5	Owens Valley, Calif., site E	37.250	118.333	NR	Jan. 1984-Oct. 1985	Duell, 1990
6	Owens Valley, Calif., site F	37.108	118.250	NR	Jan. 1984-Oct. 1985	Duell, 1990
7	Owens Valley, Calif., site G	36.983	118.225	NR	Jan. 1984-Oct. 1985	Duell, 1990
8	Owens Valley, Calif., site I	36.842	118.183	NR	Jan. 1984-Oct. 1985	Duell, 1990
9	Owens Valley, Calif., site L	36.783	118.183	NR	Jan. 1984-Oct. 1985	Duell, 1990
10	Ash Meadows, Nev., site 1	36.482	116.332	2,255	March-Dec. 1994	Nichols and others, 1997
11	Ash Meadows, Nev., site 2	36.482	116.335	2,252	March-Dec. 1994	Nichols and others, 1997
12	Railroad Valley, Nev.	38.503	115.769	4,757	June 1992-Dec. 1994	Nichols, 1994

Table A2. Vegetation characteristics and depth to ground water at field sites, Great Basin study areas

[E, Estimated.]

Site (fig. A1, table A1)	Most common plant types	Plant density (*d*)	Maximum plant leaf area index (*LAIp*)	Minimum depth to ground water (feet below land surface)
1	Greasewood, saltbush, sagebrush [1]	0.17	2.7	8.9
2	Greasewood, rabbitbrush	.21	3.4	5.9
3	Alkali sacaton, russian thistle, bassia, saltgrass [2]	.42	2.0E	10.5
4	Saltgrass, rabbitbrush, alkali sacaton, saltbush, greasewood [2]	.35	1.0E	10.2
5	Rabbitbrush, alkali sacaton, mormon tea, sagebrush, saltgrass, greasewood [2]	.26	1.8E	10.2
6	Saltgrass, greasewood, alkali sacaton, saltbush [2]	.24	1.5E	7.9
7	Saltgrass, alkali sacaton, rabbitbrush, greasewood [2]	.33	1.9E	7.2
8	Saltbush, alkali sacaton, rabbitbrush, greasewood [2]	.50	1.8E	4.6
9	Saltgrass, alkali sacaton, wiregrass[2]	[3].73	2.6	.0
10	Saltgrass	.60	2.8	1.6
11	Saltgrass, wiregrass	.95	3.5	.0
12	Greasewood, saltbush	.13	1.4	5.9

[1] Perched ground water. Depth to water table is 20 feet.

[2] From Duell (1990).

[3] From Groeneveld and Warren (1992).

density and plant species reported by Duell (1990) and limited *LAIp* values given by Groeneveld and Warren (1992). The leaf area index, *LAI*, is given by

$$LAI = d\overline{(LAIp)_{max}} \tag{1}$$

where d is measured plant density, and

$\overline{(LAIp)_{max}}$ is the weighted-average maximum leaf area index of shrubs along the measured plant-density transect.

The leaf area index, *LAI*, then was normalized by dividing by 4, the assumed maximum value for *LAI*; the resulting index, which is referred to as plant cover, *Cp*, is given by

$$Cp = \frac{LAI}{4}. \tag{2}$$

Studies at field sites 1 and 2 in 1989 and 1991 followed several years of drought. Winter precipitation in western Nevada was sparse, and was evapotranspired by early to mid-May. The measured mean daily evapotranspiration from late May or early June to early September at each site was assumed therefore to represent mean daily ground-water evapotranspiration from phreatophyte shrubs for May through September. Field observations have shown that summer convective-storm precipitation is evapotranspired within 5 to 7 days. Consequently, evapotranspiration for periods of 5 to 7 days following convective storms was not included in the estimation of mean daily ground-water evapotranspiration.

Studies in 1994 at field sites 10 and 11 in the Ash Meadows area in southern Nevada also followed an extended dry period. Precipitation was not measured at the field sites; the nearest U.S. Weather Service stations are about 10 mi north and about 15 mi southeast of the study sites. On the basis of data from these stations, 0.65 inch of precipitation fell at locations near the study sites during the last 3 months of 1993 and as much as 1.44 inches fell during January and February 1994. No precipitation fell at field sites 10 and 11 during 1994 after February. Therefore, evapotranspiration from May 1 through September 30 at these sites was assumed to be derived entirely from ground water. The October through April evapotranspiration from the Ash Meadows sites was calculated using the January through April and October through December 1994 data; January and February precipitation was subtracted from the October to April total before calculating the mean daily evapotranspiration.

Duell (1990) measured precipitation only at field sites 4, 6, and 9 (Duell's sites C, F, and L; Duell, 1990). Evapotranspiration from Owens Valley field sites at which precipitation data were not collected have been corrected by subtracting precipitation recorded at the nearest field site. Evapotranspiration at field sites 3 and 5 (Duell's sites A and E) has been corrected by subtracting precipitation measured at field site 4. Evapotranspiration at field site 7 (Duell's site G) has been corrected by subtracting precipitation measured at field site 6, and evapotranspiration from field site 8 (Duell's site J) has been corrected by subtracting precipitation measured at field site 9.

Duell (1990, p. E25) presented data for January 1984 through October 1985. For the present study, May through September evapotranspiration (table A3) is the mean of May-September 1984 and May-September 1985 evapotranspiration reported by Duell (1990). October through April evapotranspiration (table A3) is for October through December 1984 and January through April 1985 (table A3).

GROUND-WATER EVAPOTRANSPIRATION FROM PHREATOPHYTE SHRUBS AND GRASSES AND FROM ASSOCIATED BARE SOIL

Evapotranspiration as a Function of Plant Cover

Measurements of May through September ground-water evapotranspiration from shrubs and saltgrass field sites in Nevada (sites 1, 2, 10, 11, and 12, fig. A2, tables A1 and A2) were the foundation for the analysis. However, because only two field sites (11 and 12) included data for October through April, data from the Owens Valley field sites were included so that the results of the analysis could extend to winter and annual estimates of ground-water evapotranspiration. May-September (153 days), October-April (212 days), and annual ground-water evapotranspiration (table A3) are plotted as a function of plant cover at each study site (figs. A2-A4). Least-squares analysis indicated the curve that best describes the data is an exponential equation of the form

$$ET = \exp\left[a + \frac{b}{Cp} + c\ \ln(Cp)\right] \tag{3}$$

where *ET* is mean daily May-September, mean daily October-April, annual mean daily, or annual total ground-water evapotranspiration.

Coefficients *a, b,* and *c* for estimating seasonal and annual ground-water evapotranspiration and the coefficient of determination, r^2, for each data set in table A3 are given in table A4.

Table A3. Seasonal and annual rates of ground-water evapotranspiration at field sites, Great Basin study areas

Site (fig. A1, table A1)	Mean daily ground-water evapotranspiration (feet per day)			Mean annual ground-water evapotranspiration (feet)
	May-September	October-April	Annual	
1	0.0054	--	--	--
2	.0080	--	--	--
3	.010	0.0021	0.0054	1.97
4	.0043	.00078	.0023	.839
5	.0049	.0022	.0033	1.20
6	.0024	.00058	.0013	.474
7	.0070	.0020	.0041	1.46
8	.010	.0024	.0055	2.04
9	.014	.0028	.0075	2.73
10	.012	.0029	.0067	2.45
11	.013	.0025	.0069	2.52
12	.0013	--	--	--

Equation 3 was selected from several equations that equally well described the evapotranspiration-plant cover relation (all equations had an $r^2 \geq 0.96$). This equation was chosen because it is equivalent to the equation used to calculate saturation vapor pressure as a function of temperature (Arya, 1988, p. 52).

Equation 3, therefore, may have a physical basis in the calculation of ET, compared to the strictly empirical relation described by the other candidate equations.

Evapotranspiration as a Function of Depth to Ground Water

The same ground-water evapotranspiration data (table A3) used in the above analysis are plotted against the depth to ground water (table A2) at each of the field sites in figures A5-A7. In all cases, the data are best described by a linear equation

$$ET = \alpha + \beta Z_w \quad \text{for } Z_w < 10\text{ft}, \qquad (1)$$

where ET is mean daily May-September, mean daily October-April, annual mean daily, or annual total ground-water evapotranspiration, and

Z_w is depth to ground water, in feet.

Coefficients α and β for estimating seasonal and annual ground-water evapotranspiration and the coefficient of determination, r^2, for each data set in table A3 are given in table A5.

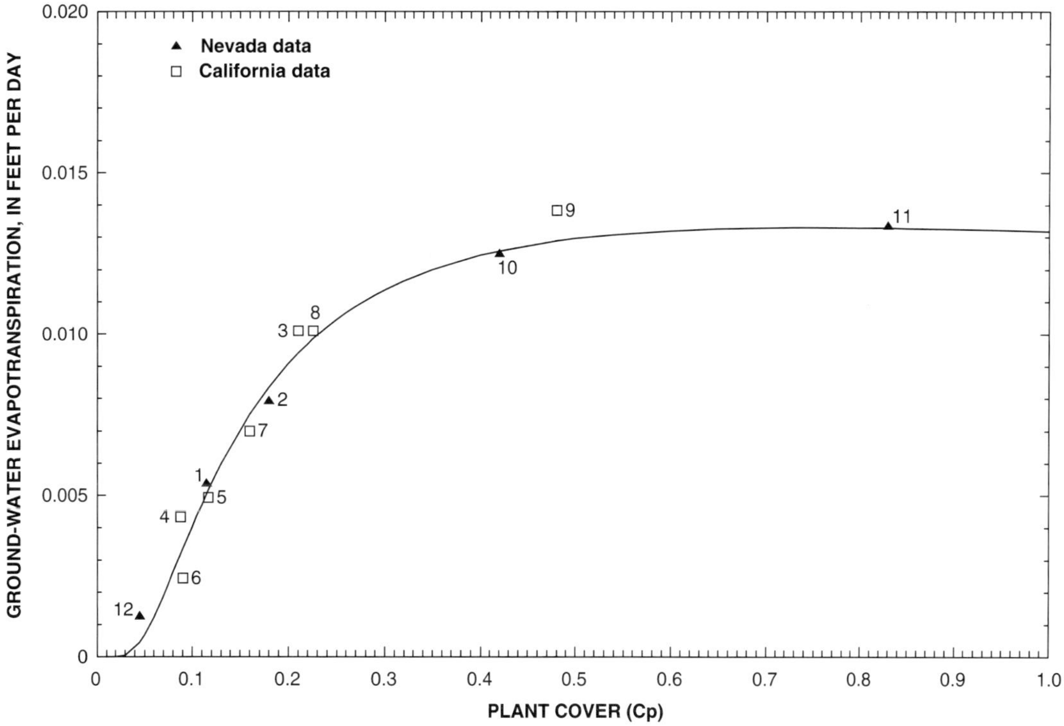

Figure A2. May through September ground-water evapotranspiration from phreatophyte shrubs and grasses and associated bare soil as related to plant cover. Numbers refer to field sites shown on figure A1 and described in table A1.

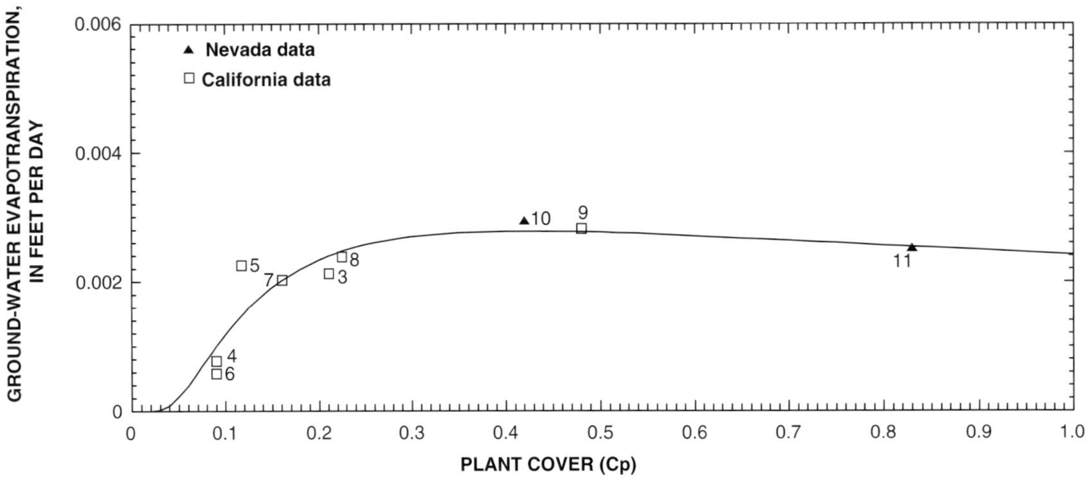

Figure A3. October through April ground-water evapotranspiration from phreatophyte shrubs and grasses and associated bare soil as related to plant cover. Numbers refer to study sites shown on figure A1 and described in table A1.

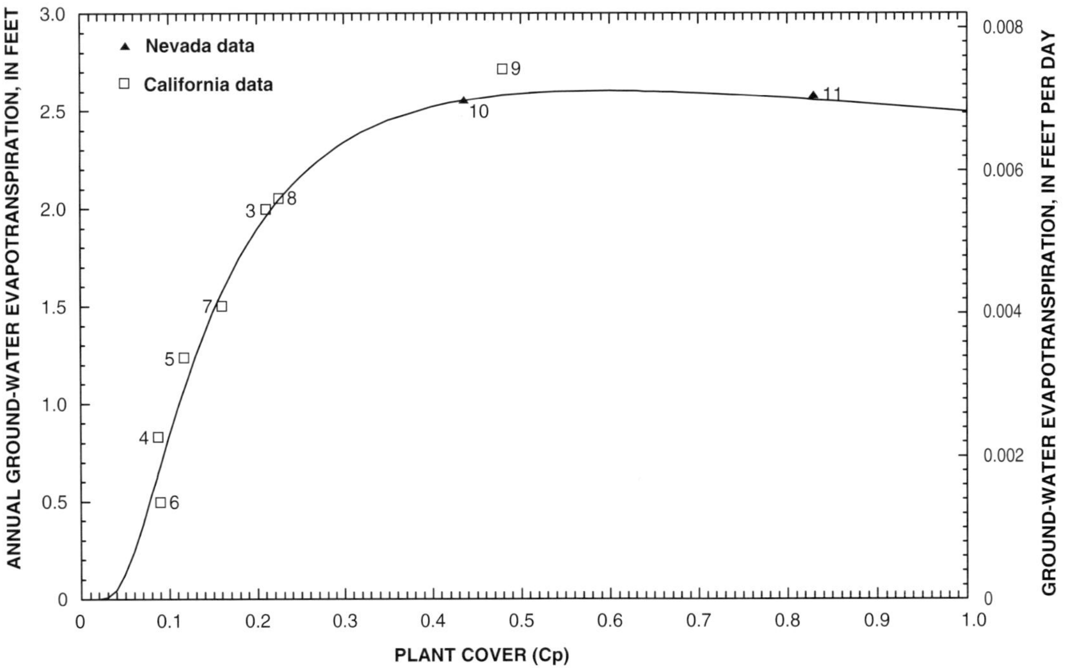

Figure A4. Annual ground-water evapotranspiration from phreatophyte shrubs and grasses and associated bare soil as related to plant cover. Numbers refer to study sites shown on figure A1 and described in table A1.

Table A4. Coefficients for equation for estimating ground-water evapotranspiration as a function of plant cover

Data set (table A3)	Coefficients *a*	*b*	*c*	r^2
May-September, feet per day	-4.13	-0.199	-0.263	0.973
October-April, feet per day	-5.82	-.203	-.483	.842
Annual, feet per day	-4.77	-.214	-.358	.975
Annual, feet	1.13	-.215	-.363	.975

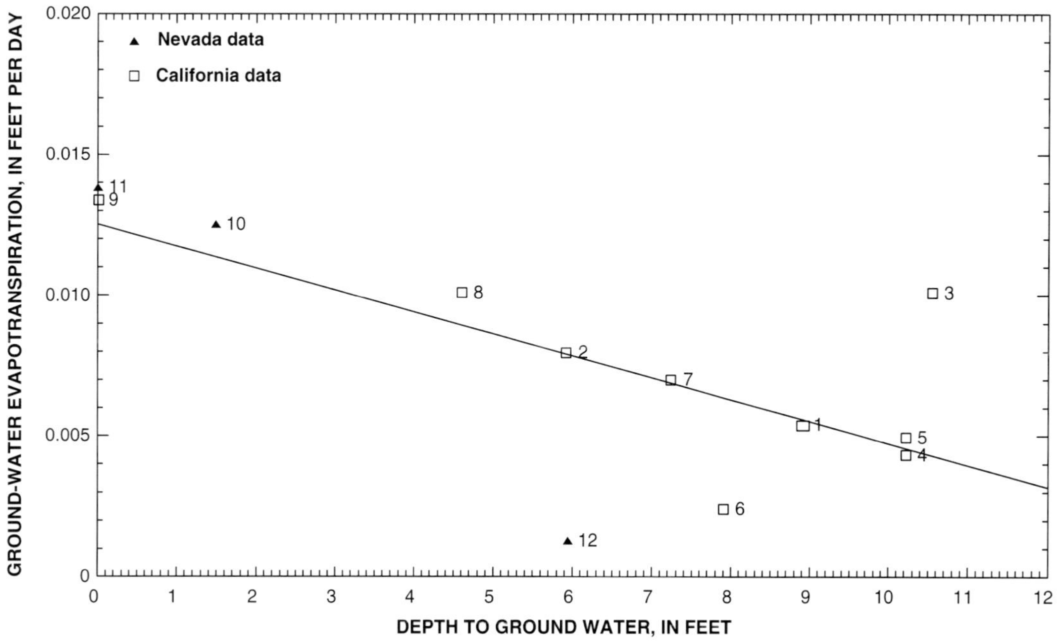

Figure A5. May through September ground-water evapotranspiration from phreatophyte shrubs and grasses and associated bare soil as related to depth to ground water. Numbers refer to field sites shown on figure A1 and described in table A1.

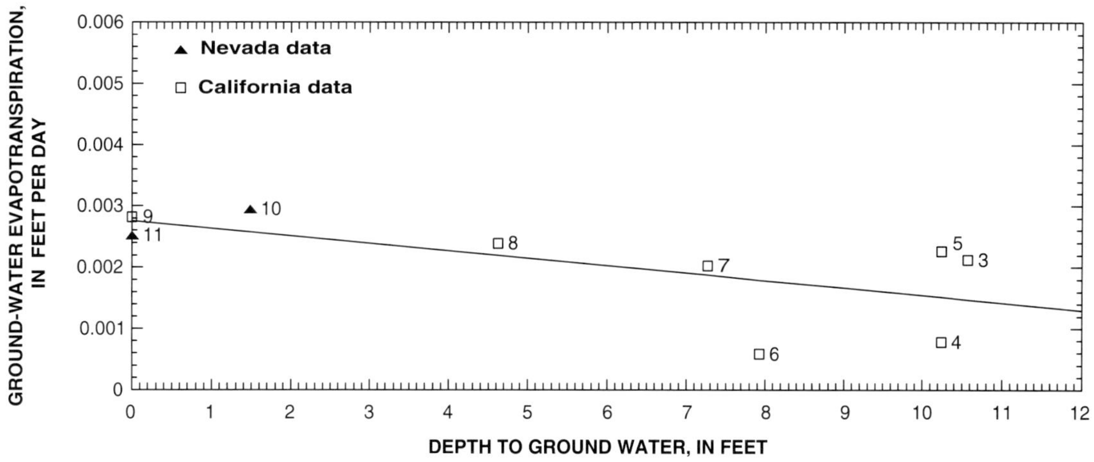

Figure A6. October through April ground-water evapotranspiration from phreatophyte shrubs and grasses and associated bare soil as related to depth to ground water. Numbers refer to field sites shown on figure A1 and described in table A1.

Table A5. Coefficients for equation for estimating ground-water evapotranspiration as a function of depth to ground water

Data set (table A3)	Coefficients		r^2
	α	β	
May-September, feet per day	0.0125	-0.00078	0.505
October-April, feet per day	.00276	-.000121	.415
Annual, feet per day	.00715	-.000394	.627
Annual, feet	2.61	-.143	.626

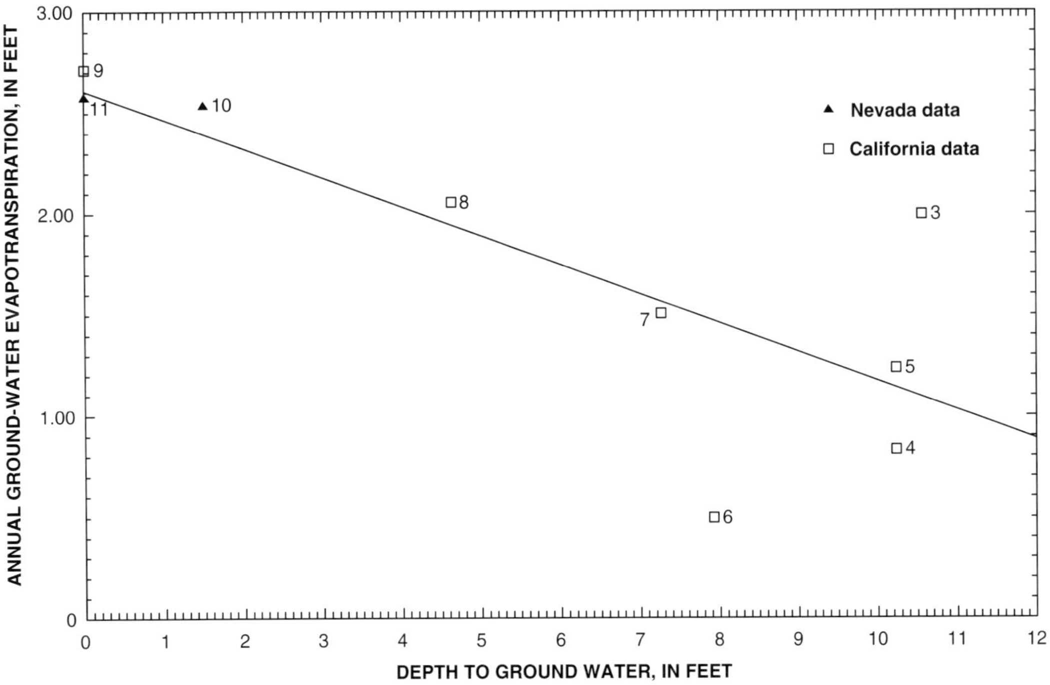

Figure A7. Annual ground-water evapotranspiration from phreatophyte shrubs and grasses and associated bare soil as related to depth to ground water. Numbers refer to field sites shown on figure A1 and described in table A1.

The correlation between ground-water evapotranspiration and depth to ground water is good except at field sites 3, 6, and 12. Duell (1990, p. E6) reported that ground-water levels were recorded at wells within 500 ft of each of his field sites. It may be that the depth to ground water recorded for field sites 3 and 6 is not representative of the depth to water beneath the sites or it may be that evapotranspiration rates are underestimated at field site 6 and overestimated at field site 3. Ground-water evapotranspiration at field site 12 is anomalously low for the depth to ground water at the site, indicating something other than depth to ground water is influencing the amount of plant cover, and hence the amount of ground-water evapotranspiration, at the site.

Relation Between Plant Cover and Depth to Ground Water

The good correlation between ground-water evapotranspiration and plant cover and between ground-water evapotranspiration and depth to ground water strongly suggests a correlation between plant cover and the depth to ground water, the assumption upon which the foregoing analysis was based. Plant cover is plotted against the annual minimum depth to ground water at each study site in figure A8. The data shown in figure A8 are best described ($r^2 = 0.837$) by an exponential equation of the form

$$Cp = \exp\left[-0.534 + \left(\frac{-0.0049}{Z_w}\right) - 0.730\ln(Z_w)\right] \quad (5)$$

$$\text{for } Z_w > 0$$

where Cp is the plant cover, in this case, that reasonably may be expected to occur for a given depth to ground water, and

Z_w is depth to ground water, in feet.

GROUND-WATER EVAPOTRANSPIRATION, PHREATOPHYTE PLANT COVER, AND DEPTH TO GROUND WATER

The good correlation between plant cover and ground-water evapotranspiration and between plant cover and depth to ground water combined with the weaker correlation between ground-water evapotranspiration and depth to ground water strongly suggest that plant cover is the major factor in determining ground-water evapotranspiration by phreatophytes in areas of shallow ground water. Even in areas of very shallow ground water, ground-water evapotranspiration is small if plant cover is sparse, such as at site 12.

The equations developed by this study provide a consistent, or uniform, method for estimating ground-water evapotranspiration from phreatophyte shrubs and grasses and from associated bare soil in the Great Basin; they are not intended to be used to estimate ground-water evapotranspiration by saltcedar, willows and other trees, or from riparian vegetation. Ground-water evapotranspiration estimated with the equations proposed in the present study includes evaporation of ground water from the water table through bare soil and transpiration of ground water by phreatophyte shrubs and grasses. Those equations proposed in the earlier study (Nichols, 1994) estimated only the ground water transpired by phreatophyte shrubs and implied that *LAI* was independent of the depth to ground water.

Mean daily and annual evapotranspiration estimated by the equations presented here may seem large in comparison to estimates suggested by earlier studies, but these values must be understood in the context of the depth to ground water for which they are appropriate (fig. A9). Annual ground-water evapotranspiration greater than about 1.15 ft/yr applies to areas where the depth to ground water is less than about 10 ft. This occurs commonly in grassy and marshy areas near springs that represent a small part of the hydrographic areas and basins throughout most of the Great Basin. Ground-water levels commonly are 15 to 25 ft deep beneath most rangeland areas with phreatophyte shrubs such as greasewood and rabbitbrush. Annual ground-water evapotranspiration in these areas will range from about 0.16 to about 0.50 ft/yr.

Correlating plant cover with depth to ground water is not entirely straightforward. Factors other than the depth to the water table influence plant cover at any given location. For example, the depth to the water table at field site 1 in the Smoke Creek Desert is 20 ft, but there is a perched saturated zone at about 9 ft. The plant cover at this site fits the proposed curve (fig. A8) better at a depth to ground water of about 10 ft than it does a depth of 20 ft. However, unknown factors at field site 12 in Railroad Valley have resulted in a plant cover at the site of only about 0.045 (4.5 percent) in an area where the depth to ground water is 5.9 ft; this plant cover is more consistent with a depth to ground water of about 30 to 35 ft. Consequently, the relation defined by equation 5 describes the plant cover that might be expected for the indicated depth to the water table in the absence of other factors that affect the actual plant cover.

The equations presented here and the data upon which they are based provide insights into the processes of ground-water evapotranspiration by phreatophyte shrubs and grasses in the Great Basin. This process is shown by the curve in figure A4. Maximum ground-water evapotranspiration occurs in areas where the plant cover ranges from about 0.50 to 0.65 (50 percent to 65 percent), and areas with a plant cover of only about 0.40 (40 percent) discharge as much ground water by evapotranspiration as areas where the plant cover is 1.00 (100 percent). These conclusions, although not necessarily intuitive, are supported by observations at field sites 10 and 11 (tables A2 and A3) in the Ash Meadows area in southern Nevada (Nichols and others, 1997). Field site 10 has a plant cover of about 0.42 (42 percent) leaving about 58 percent of the area as bare soil. The minimum depth to ground water at the site was about 1.6 ft. Field site 11 had a plant cover of about 0.83 (83 percent) with ground water at or slightly above land surface during winter months. In spite of the large difference in the plant cover between the two field sites, the annual ground-water evapotranspiration from each site is similar. A comparison of daily ground-water evapotranspiration for the two sites (Nichols and others, 1997) shows that ground-water evapotranspiration increased more rapidly in the spring at field site 10 than at field site 11. The saltgrass at field site 10 was a vigorous green by mid-May; at field site 11, it did not become green and show evidence of growth until late May, although the wire-grass became green much earlier. The difference in ground-water evapotranspiration at the two field sites during this time reflects the greater evaporation of shallow ground water from bare soil and somewhat greater transpiration from saltgrass at field site 10. Dormant vegetation at field site 11 appears to have formed an insulating cover that shaded and reduced evaporation from the underlying soil during the spring and early summer. By July and August, the greater plant cover at field site 11 is reflected in greater daily and monthly ground-water evapotranspiration (Nichols and others, 1997). Ground-water evapotranspiration at field site 10 once more exceeded that at field site 11 from October to the end of the year, again reflecting the shading of soil by dormant vegetation at field site 11.

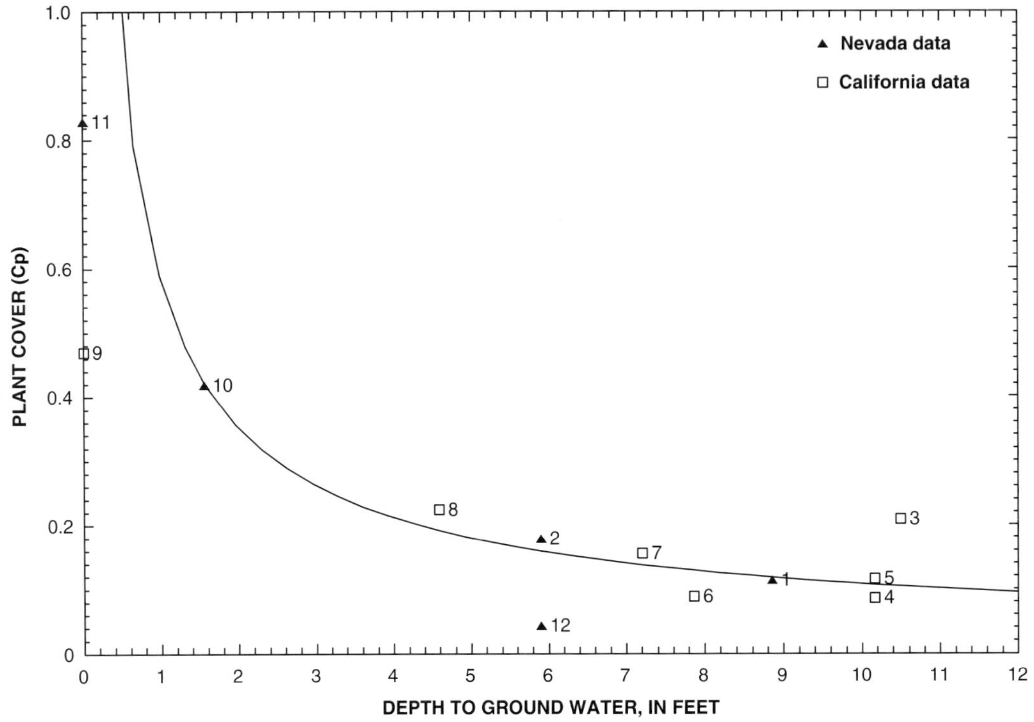

Figure A8. Relation between plant cover and depth to ground water. Numbers refer to field sites shown on figure A1 and described in table A1.

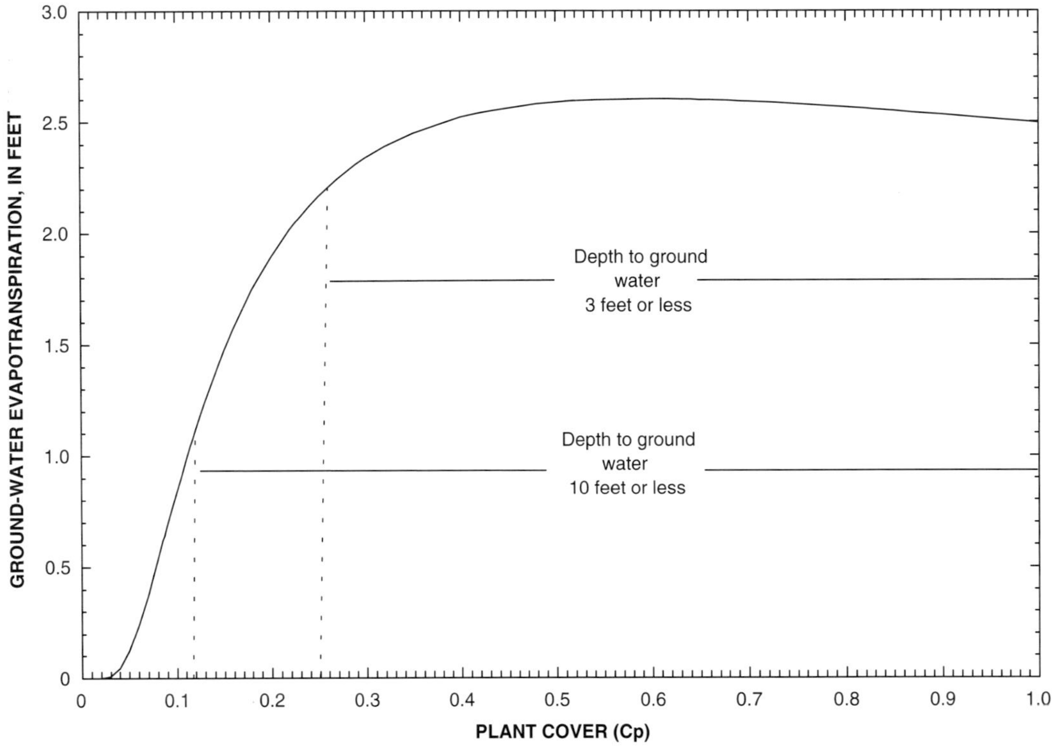

Figure A9. Annual ground-water evapotranspiration from phreatophyte shrubs and grasses and associated bare soil and the approximate depth to ground water.

Winter (October through April) ground-water evapotranspiration is considered small to insignificant, and in areas where the depth to ground water is greater than 10 to 15 ft, this is true. However, in areas where the depth to ground water is 3 ft or less, October to April ground-water evapotranspiration may be as much as 0.6 ft and account for up to a quarter of the annual ground-water evapotranspiration. Most of this is evaporation from bare soil, because plants are dormant during much of this time. Some transpiration of ground water will occur, however, because phreatophyte shrubs remain green well into October and, on occasion, into November. During periods in the winter when soils near the surface are frozen, soil moisture and soil-water vapor derived from ground water continue to move toward the surface in response to soil-water tension gradients, and accumulate. When the frozen surface soils thaw, this ground water-derived soil moisture will be evapotranspired.

SUMMARY AND CONCLUSIONS

The equations presented here describe a functional relation: (1) between seasonal and annual ground-water evapotranspiration and plant cover, and (2) between seasonal and annual ground-water evapotranspiration and the depth to shallow ground water. They are an alternative to equations for estimating ground-water evapotranspiration by phreatophyte shrubs in the Great Basin proposed previously by Nichols (1994). These equations are not intended to be used to estimate ground-water evapotranspiration by saltcedar, willows, cottonwood trees, or other riparian vegetation.

The proposed equations for estimating ground-water evapotranspiration as a function of plant cover are appropriate for estimating regional-scale ground-water evapotranspiration from remotely sensed vegetation index data, provided correlation can be made between plant cover used in this analysis and satellite-data-derived vegetation index. Equations that describe ground-water evapotranspiration as a function of the depth to ground water are appropriate for use in numerical models of ground-water flow and for other applications where the depth to ground water is known.

REFERENCES CITED

Arya, S.P., 1988, Introduction to micrometeorology: San Diego, Calif., Academic Press, International Geophysics Series, v. 42, 307 p.

Blaney, H.F., Ewing, P.A., Israelson, O.W., Rohwer, C., and Scobey, F.C., 1938, Regional Planning, Part IV - The Rio Grande Joint Investigation in the upper Rio Grande Basin in Colo., N. Mex., and Tex., 1936-1937, Part III - Water Utilization: United States Bureau of Agricultural Engineering Report, National Resources Committee, v. 1, p. 293-427.

Blaney, H.F., Taylor, C.A., and Young, A.A., 1930, Rainfall penetration and consumptive use of water in Santa Ana River Valley and coastal plain: California Department of Public Works, Water Resources Division Bulletin 33, 162 p.

Blaney, H.F., Taylor, C.A., Young, A.A., and Nickle, H.G., 1933, Water losses under natural conditions in wet areas in southern California, Part I, Consumptive use by native plants growing in moist areas of southern California: California Department of Public Works, Water Resources Division Bulletin 44, p. 19-139.

Businger, J.A., Miyake, M., Dyer, A.J., and Bradley, E.F., 1967, On the direct determination of the turbulent heat flux near the ground: Journal of Applied Meteorology, v. 6, p. 1025-1032.

Duell, L.F.W., Jr., 1990, Estimates of evapotranspiration in alkaline scrub and meadow communities of Owens Valley, California, using the Bowen-ratio, eddy-correlation, and Penman-Monteith methods: U.S. Geological Survey Water-Supply Paper 2370-E, 39 p.

Gatewood, J.S., Robinson, T.W., Colby, B.R., Hem, J.D., and Halpenny, L.C., 1950, Use of water by bottom-land vegetation in lower Safford Valley, Arizona: U.S. Geological Survey Water-Supply Paper 1103, 210 p.

Gay, L.W., and Fritschen, L.J., 1979, An energy balance budget analysis of water use by saltcedar: Water Resources Research, v. 15, p. 1589-1592.

Groeneveld, D.P., and Warren, D.C., 1992, Total transpiration from land areas estimated from hand-held porometer measurements, *in* Wilson, D.H., Reginato, R.J., and Hollett, K.J., eds., Evapotranspiration measurements of native vegetation, Owens Valley, California, June 1986: U.S. Geological Survey Water-Resources Investigations Report 91-4159, p. 49-59.

Harrill, J.R., Gates, J.S., and Thomas, J.M., 1988, Major ground-water flow systems in the Great Basin region of Nevada, Utah, and adjacent states: U.S. Geological Survey Hydrologic Investigations Atlas HA-694-C, 2 sheets.

Lee, C.H., 1912, An intensive study of the water resources of a part of the Owens Valley, California: U.S. Geological Survey Water-Supply Paper 294, 135 p.

Malek, E., Bingham, G.B., and McCurdy, G.D., 1990, Evapotranspiration from the margin and moist playa of a closed desert valley: Journal of Hydrology, v. 120, p. 15-34.

Monteith, J.L., and Unsworth, M.H., 1990, Principles of environmental physics (2d ed.): New York, Edward Arnold, 291 p.

Mozingo, N.H., 1987, Shrubs of the Great Basin: Reno, University of Nevada Press, 342 p.

Nichols, W.D., 1994, Groundwater discharge by phreatophyte shrubs in the Great Basin as related to depth to groundwater: Water Resources Research, v. 30, no.12, p. 3265-3274.

Nichols, W.D., Laczniak, R.J., DeMeo, G.A., and Rapp, T.R., 1997, Estimated ground-water discharge by evapotranspiration, Ash Meadows area, Nye County, Nevada, 1994: U.S. Geological Survey Water-Resources Investigation Report 97-4025, 13 p.

Robinson, T.W., 1958, Phreatophytes: U.S. Geological Survey Water-Supply Paper 1423, 84 p.

———1970, Evapotranspiration by woody phreatophytes in the Humboldt River valley near Winnemucca, Nevada: U.S. Geological Survey Professional Paper 491-D, 41 p.

Smith, J.L., Reece, B.D., and Medina, R.L., 2000, Data sets and related information for regional ground-water evapotranspiration in eastern Nevada: U.S. Geological Survey Open-File Report 99-242, CD-ROM.

Tanner, C.B., 1960, Energy balance approach to evapotranspiration from crops: Soil Science Society of America Proceedings, 24, p. 1-9.

Weeks, E.P., Weaver, H.L., Campbell, G.S., and Tanner, B.D., 1987, Water use by saltcedar and by replacement vegetation in the Pecos River floodplain between Acme and Artesia, New Mexico: U.S. Geological Survey Professional Paper 491-G, 33 p.

White, W.N., 1932, A method of estimating ground-water supplies based on discharge by plants and evaporation from soil—Results of investigations in Escalante Valley, Utah, *in* Contributions to the hydrology of the United States 1932: U.S. Geological Survey Water-Supply Paper 659, p. 1-105.

Young, A.A., and Blaney, H.F., 1942, Use of water by native vegetation: California Department of Public Works, Division of Water Resources Bulletin 50, 160 p.

Chapter B. Estimating Regional Ground-Water Evapotranspiration From Phreatophytes, Great Basin, Nevada

By William D. Nichols, J. LaRue Smith, *and* Brian D. Reece

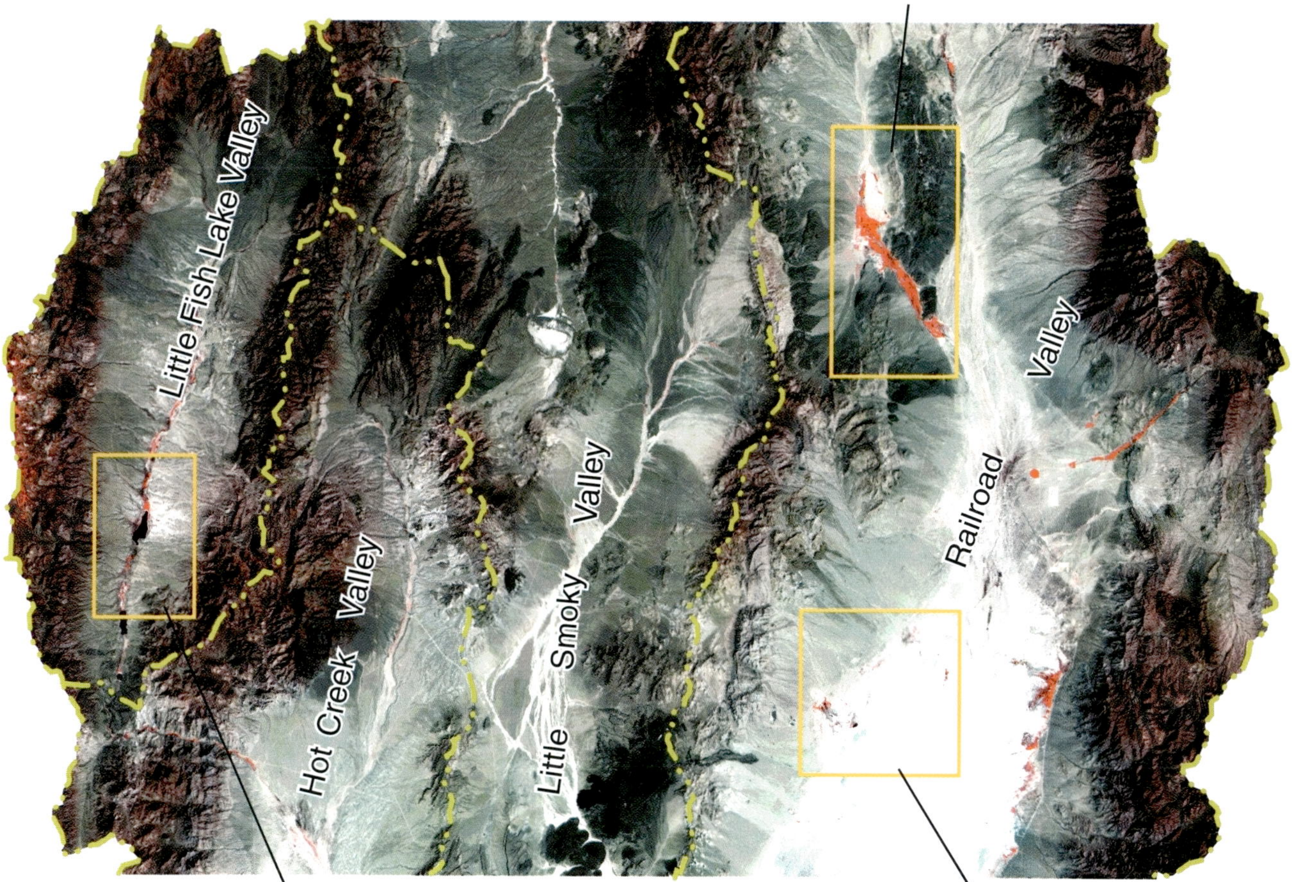

Title-page photograph. Landsat Thematic Mapper color-infrared image, acquired August 13, 1985, of part of Nye County, Nevada. Bright red colors generally are broadleaf, healthy vegetation. Reddish brown colors generally are needle-leaf type vegetation. White colors are bare soil and low-density vegetation cover. Light to dark greenish colors are rangeland. Water is dark blue to black. Rectangles identify field-study areas discussed in this chapter. Yellow dash-and-dot lines are topographic boundaries between valleys. Map scale, about 10 miles per inch.

CONTENTS

FIGURES

TABLES

Chapter B. Estimating Regional Ground-Water Evapotranspiration From Phreatophytes, Great Basin, Nevada

By William D. Nichols, J. LaRue Smith, and Brian D. Reece

ABSTRACT

Previous ground-water studies estimated ground-water evapotranspiration by phreatophytes and bare soil in Nevada on the basis of results of field studies published in 1912 and 1932. More recent studies of evapotranspiration by rangeland phreatophytes, using micrometeorological methods as discussed in Chapter A of this report, provide new data on which to base estimates of ground-water evapotranspiration. An approach correlating ground-water evapotranspiration with plant cover is used in conjunction with a modified soil-adjusted vegetation index derived from Landsat data to develop a method for estimating the magnitude and distribution of ground-water evapotranspiration at a regional scale. Large areas of phreatophytes near Duckwater and Lockes in Railroad Valley are believed to subsist on ground water discharged from nearby regional springs. Ground-water evapotranspiration by the Duckwater phreatophytes of about 11,500 acre-feet estimated by the method described in this report compares well with measured discharge of about 13,500 acre-feet from the springs near Duckwater. Measured discharge from springs near Lockes was about 2,400 acre-feet; estimated ground-water evapotranspiration using the proposed method was about 2,450 acre-feet.

INTRODUCTION

Evapotranspiration by phreatophyte shrubs and grasses and evaporation from bare soil are the principal mechanisms of ground-water discharge from the valleys of the Great Basin. Previous estimates of ground-water evapotranspiration by phreatophytes in Nevada were based on studies by Lee (1912) in Owens Valley, Calif.; White (1932) in the Escalante Desert, Utah; and, to a lesser extent, Robinson (1970) near Winnemucca, Nev. (fig. B1). Estimates of ground-water evapotranspiration that were based on these studies have been used to estimate ground-water budgets for most of the valleys and hydrographic areas of Nevada (Rush, 1968). The estimated ground-water budgets, in turn, are used to allocate ground-water resources by the Nevada State Engineer. Any attempt to improve upon the estimated budgets requires an improvement of the data upon which the estimates were based, specifically estimates of ground-water evapotranspiration.

Small, battery-powered data loggers and micrometeorological instruments suitable for completing energy budget evapotranspiration studies at remote locations became available in the early 1980's. The U.S. Geological Survey began a series of field studies in 1988 (Nichols, 1992, 1993, 1994; Nichols and Rapp, 1996; Nichols and others, 1997) to measure ground-water evapotranspiration from phreatophytes typical to the Great Basin. These studies provided revised evapotranspiration rates and a better understanding of evapotranspiration processes from which ground-water evapotranspiration estimates could be reevaluated. Correlation between plant cover and ground-water evapotranspiration (as discussed in Chapter A of this report) provided equations from which to estimate regional ground-water evapotranspiration using Landsat data.

This chapter describes a method for estimating regional ground-water evapotranspiration by phreatophyte shrubs and grasses and associated bare soil in the Great Basin as a function of plant cover determined from a vegetation index derived from Landsat data.

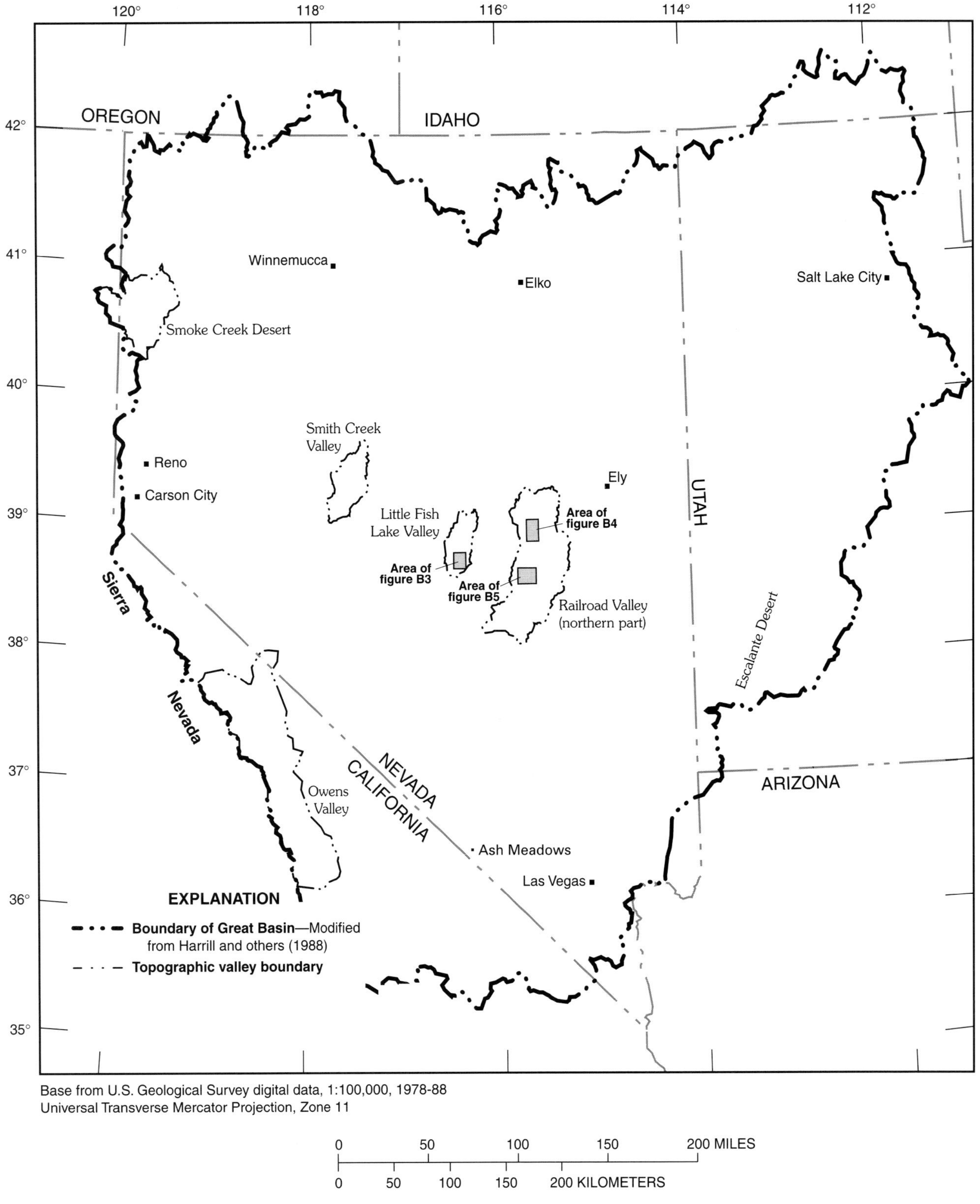

Figure B1. Location of plant cover study areas, Great Basin, Nevada.

ESTIMATING REGIONAL GROUND-WATER EVAPOTRANSPIRATION BY PHREATOPHYTES

The early studies of Lee (1912) and White (1932), and the somewhat later study of Robinson (1970), provided preliminary data for estimating ground-water evapotranspiration from saltgrass and phreatophyte shrubs in the Great Basin. The results of these studies, however, were difficult to apply and transfer to other areas because the studies did not correlate the estimated ground-water evapotranspiration with uniquely defining criteria, such as leaf area index.

More recent studies (Carman, 1989; Duell, 1990; Malek and others, 1990; Nichols, 1992; 1993) using micrometeorological methods to measure evapotranspiration from rangeland phreatophytes also did not include information needed to transfer the site-specific results to areas with different vegetation and ground-water conditions. Transferability problems can in part be attributed to the difficulty of collecting evapotranspiration data at a sufficient number of rangeland locations and over a long enough time to begin addressing issues of similarity and variability of results among sites.

A relation between ground-water evapotranspiration from phreatophytes and the depth to ground water has been assumed and, to some extent, demonstrated, but until recently only Lee (1912) had proposed equations to describe this relation for saltgrass. Using field measurements and the results of an energy-combination model to separate soil evaporation and plant transpiration, Nichols (1994) proposed a functional relation between phreatophyte-shrub ground-water transpiration and shrub density, shrub leaf area index, and the depth to ground water. Subsequently (Chapter A), the analysis was extended to develop a functional relation between plant cover and ground-water evapotranspiration from phreatophyte shrubs and saltgrass and associated bare soil. The following discussion summarizes that analysis, which is described in detail elsewhere (see Chapter A).

GROUND-WATER EVAPOTRANSPIRATION AS A FUNCTION OF PLANT COVER

Energy-budget studies of ground-water evapotranspiration from native rangeland were done in the Smoke Creek Desert, Smith Creek Valley, Railroad Valley, and the Ash Meadows area, Nev. (fig. B1; Nichols, 1994; Nichols and others, 1997). Data from these sites and from seven sites in Owens Valley, Calif. (Duell, 1990), were used to develop a correlation between ground-water evapotranspiration and plant cover. Site descriptions and data used in the analysis are given in Chapter A of this report.

The analysis in Chapter A concluded that ground-water evapotranspiration from saltgrass/meadow grass, phreatophyte shrubs, and the associated bare soil could be estimated as a function of plant cover determined from plant density (*d*) and leaf area index (*LAI*). Chapter A discusses these two variables, and gives an equation that relates the two in terms of the shrub leaf area index (*LAIp*), which is the leaf area per unit ground area for individual shrubs along a measured plant-density transect. The equations are

$$LAI = d\overline{(LAIp)_{\max}} \quad (1)$$

and

$$C_p = \frac{LAI}{4} \quad (2)$$

where d, $\overline{(LAIp)_{\max}}$, and Cp are defined in Chapter A.

Measured evapotranspiration at each field site was corrected by subtracting precipitation that occurred during the measurement period. The resulting value for ground-water evapotranspiration includes evaporation of ground water from any bare soil at the field site, but does not include evaporation of soil moisture from recent precipitation. Least-squares analysis (Chapter A) indicated the curve that best described the data is an exponential equation of the form

$$ET = \exp\left[a + \frac{b}{Cp} + c\ln(Cp)\right] \quad (3)$$

where ET is mean daily May-September, mean daily October-April, annual mean daily, or annual total ground-water evapotranspiration;
Cp is plant cover; and
a, b, and c are coefficients for estimating seasonal and annual ground-water evapotranspiration.

Values for the three coefficients and for r^2, the coefficient of determination for each time interval, are given in table A4.

Seasonal and annual ground-water evapotranspiration estimated by these equations are difficult to compare with estimates from earlier studies (Lee, 1912; White, 1932; Robinson, 1970) because the earlier studies did not relate evapotranspiration rates to the leaf area index or to plant cover. Similarly, estimates that are based on these equations cannot be compared directly to estimates reported by previous ground-water resources studies in Nevada because these earlier reports provided only a qualitative description of plant cover such as sparse, moderate, moderately dense, or dense. Estimates made with equation 3 and those made in previous studies can be compared only at the regional scale of previous estimates.

Estimating Plant Cover at Regional Scales

Satellite data have been used for several decades to describe land-surface cover in many ways. Many methods have been developed for characterizing vegetation using these data, especially data from Landsat satellites. Some methods, such as those using false-color IR (infrared) images, provide a qualitative measure of vegetation that readily distinguishes healthy from stressed vegetation. Land-cover classification methods commonly provide descriptive information on the type of land cover at increasing levels of detail. The level of detail for a multilevel land-cover classification of rangeland areas might be from rangeland at the first level to shrub and brushland at the second level to sagebrush prairies at the third level (Sabins, 1987). Finally, several indices have been determined over the years from Landsat data as a means of quantitatively describing vegetation. These indices provide regional plant-cover information that is appropriate for use in the equations proposed by Nichols (see Chapter A).

Corrections and Calibration of Landsat Data

Landsat is an unmanned earth-orbiting satellite system, the first of which was launched in 1972. The satellite imagery used in this study was acquired by Landsat 5 launched in 1984. Landsat satellites are in a sun-synchronous orbit with the south-bound segment of the orbit during daylight and the north-bound segment at night. These satellites collect reflectance data continuously along orbital paths that are repeated every 16 days. The data are used to create images that are subdivided into scenes that cover about 115 mi by 115 mi on the land surface. The scenes consist of picture elements, or pixels, that have a resolution of about 90 ft by 90 ft.

Landsat includes a thematic mapper (TM) radiometer that measures visible radiation and reflected and thermal IR in seven wavelength bands ranging from 0.45 to 12.5 μm (micrometers) and a multispectral scanner (MSS) radiometer that measures visible radiation and reflected IR in four wavelength bands ranging from 0.5 to 1.1 μm. The Landsat data of interest for this study are in TM spectral bands 3 and 4, which include visible reflectance in the red wavelengths of 0.63 – 0.69 μm (band 3) and non-visible reflectance in the near-infrared (NIR) wavelengths of 0.76 – 0.90 μm (band 4). Data in these wavelength bands are used to calculate different vegetation indices. More detailed descriptions of the Landsat satellites and the data collected by them can be obtained from Sabin (1987) or other standard sources on remote sensing.

The energy from the earth's surface measured by Landsat instruments is recorded as a digital number (DN) that ranges from 0 to 255. Variations in DN across a satellite image are caused primarily by variations in surface reflectance. However, the measured energy also is affected by such factors as scene illumination, atmospheric conditions, viewing geometry, and instrument response characteristics (Lillisand and Kiefer, 1987). The TM data used for this study were acquired on June 10, 1985. An effort was made to obtain TM data during cloud-free days in late June when plant leaf area index was expected to be at a maximum. Reflectance values were corrected for the above effects so that any subsequent calculations using the vegetation indices would not be biased by factors unique to the June 10, 1985, data.

Corrections to the TM data for atmospheric effects were made using a method proposed by Chavez (1989). This technique provides for the selection of a relative atmospheric-scattering model and predicted haze values from the image data. Sensor and astronomical corrections also were applied.

Calibration of TM data and conversion to percent reflectance using this technique makes several assumptions. The procedure converts the DN recorded by the radiometer to a radiance value by correcting for the gain and offset of the sensor. Top-of-the-atmosphere radiance values are then converted to ground reflectance values using the astronomical factors of sun angle, the distance between the earth and the sun at the time of data acquisition, and the exoatmospheric spectral irradiance. Atmospheric influences include haze and atmospheric attenuation. Atmospheric haze is solar radiation that is scattered by the atmosphere into the radiometer, and is an additive component of measured radiation. Atmospheric attenuation, a mutiplicative component, is the decrease in the intensity of reflected sunlight as it travels through the atmosphere.

Attenuation is difficult to determine, and was not accounted for when making corrections to the data used in this study. Attenuation by large concentrations of water vapor in the atmosphere may affect the near infrared portion of the spectrum, but this effect is expected to be minimal in the semi-arid environment of Nevada. Atmospheric conditions may change over a region as large as that shown in a Landsat image or from valley to valley within a smaller region; however, it was assumed that the atmosphere was constant across the image.

Satellite Data and Vegetation Indices

Two of the most commonly used vegetation indices derived from Landsat data are the normalized-difference vegetation index (*NDVI*), which is given by

$$NDVI = \frac{\rho_{NIR} - \rho_{red}}{\rho_{NIR} + \rho_{red}} \quad , \tag{4}$$

and the perpendicular vegetation index (*PVI*), which is given by

$$PVI = \alpha\rho_{NIR} - \beta\rho_{red} \quad , \tag{5}$$

where ρ is the reflectance in the near-infrared (NIR) or red bands, and
α and β are soil-line variables (Qi and others, 1994).

The soil line is determined by the ratio of ρ_{NIR} to ρ_{red} for bare soil. A functionally equivalent form of the *PVI* is the weighted-difference vegetation index (*WDVI*), which is given by (Qi and others, 1994)

$$WDVI = \rho_{NIR} - \gamma\rho_{red} \quad , \tag{6}$$

where γ is the slope of the soil line.

Studies summarized by Huete (1988) noted that these indices are influenced by soil background conditions in areas with partial and sparse plant canopy conditions. Huete (1988) proposed a soil-adjusted vegetation index to compensate for the influence of soil background effects. This index included a soil-adjustment factor that varied between 0 and 1, but was constant for all reflectance values. Qi and others (1994) proposed a modified soil-adjusted vegetation index (*MSAVI*) that included an automatically determined soil-adjustment factor that may differ for every pixel. The *MSAVI* is given by

$$MSAVI = \frac{\rho_{NIR} - \rho_{red}}{\rho_{NIR} + \rho_{red} + L}(1 + L) \quad , \tag{7}$$

where L is the soil adjustment factor and is given by

$$L = 1 - 2\gamma(NDVI)(WDVI) \quad , \tag{8}$$

where γ is the slope of the soil line as in equation 6 and is equal to 1.06 (Qi and others, 1994).

The *MSAVI* is the index used in the present study.

Estimating Plant Cover From Vegetation Index Data

Few data are available with which to correlate plant cover and *MSAVI*. Qi and others (1994, p. 124) present data for plant cover from 20 percent (Cp = 0.2) to about 97 percent (Cp=0.97) and indicate a linear relation between *MSAVI* and plant cover over this range. At plant covers less than about 15 percent (Cp = 0.15), soil-noise influences become more significant despite the built-in soil-adjustment factor. However, these are the plant cover conditions that are most important throughout Nevada and the Great Basin where areas of sparse vegetation extend to hundreds of thousands of acres with annual ground-water evapotranspiration of tens to hundreds of thousands of acre-feet.

Plant cover data were collected in Little Fish Lake and Railroad Valleys in 1995 (fig. B1). Landsat data for June 1995 were not available at the time the study was completed. Measured plant cover was correlated with *MSAVI* values derived from June 1985 Landsat data. Climatic conditions during 1984-85 were similar to those in 1994-95 and it was assumed that plant cover conditions also were similar. Subsequent analysis suggests plant cover may change by several percent from one year to the next (for example, from 0.07 to 0.09 or from 0.15 to 0.12). This interannnual variability has not materially affected the correlation between field measurements and *MSAVI* determined by the present study. However, this assumption should be verified by additional analyses using contemporaneously collected field and satellite data.

Measured plant cover (fig. B2) ranged from less than 5 to 20 percent (Cp = 0.05 to 0.20), and fall within the lower end of the linear relation shown by Qi and others (1994, p. 124). Examination of *MSAVI* values for playas and other known areas of bare soil in the study area suggest that bare soil may exhibit *MSAVI* values of as much as 0.065 to 0.070. This is approximately the magnitude of the soil noise effect indicated by Qi and others (1994). From this, it was concluded that an *MSAVI* of 0.070 or less represented zero percent plant cover (Cp = 0.0). For *MSAVI* less than 0.16,

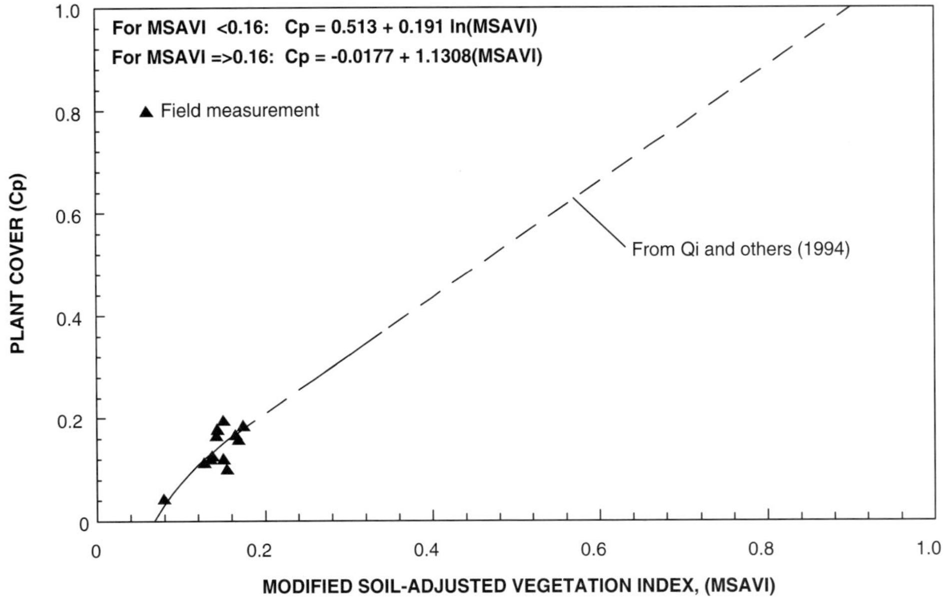

Figure B2. Correlation between modified soil-adjusted vegetation index, *MSAVI*, and plant cover, *Cp*.

the relation between *MSAVI* and plant cover was determined to be best described by a logarithmic equation ($r^2 = 0.84$)

$$Cp = 0.5130 + 0.1910\ln(MSAVI) \quad . \qquad (9)$$

For *MSAVI* equal to or greater than 0.16, the approximate relation shown by Qi and others (1994) was used (fig. B2). This relation was determined from the value of *Cp* given by equation 9 for *MSAVI* = 0.16 and the value of *MSAVI* (about 0.90) for *Cp* = 1.0 of the relation shown by Qi and others (1994, p. 124). The linear equation determined from these values is

$$Cp = -0.0177 + 1.1308(MSAVI) \quad . \qquad (10)$$

Estimating Regional Ground-Water Evapotranspiration

Data from the Landsat scene covering the area of interest were geocorrected and corrected for atmospheric effects. *NDVI*, *WDVI*, and *MSAVI* were calculated for each pixel of the scene using equations 4, 6, 7, and 8. *MSAVI* values were converted to plant cover values with equations 9 and 10. Plant cover values for individual pixels were assigned to plant-cover zones corresponding to bare soil, less than 10 percent cover, 10 to less than 20, 20 to less than 35, 35 to less than 50, and 50 percent and greater. Each zone was assigned a color used to display the plant cover distribution on the Landsat scene image.

The resulting plant-cover distribution applies to vegetation over the entire Landsat scene. To restrict an analysis only to phreatophyte areas, it is necessary to define the boundaries of phreatophytes in the area of interest. Field mapping of phreatophyte area boundaries for this study was done on U.S. Geological Survey 1:24,000 topographic quadrangle maps. The outer boundary of the phreatophyte area was assumed to represent the limit beyond which the depth to ground water was too great for phreatophytes to grow and within which the depth to ground water was sufficiently shallow that any plants growing there were using ground water. The field-verified phreatophyte boundary was transferred to the Landsat image of the study area to define the area for which ground-water evapotranspiration was calculated from plant cover. *MSAVI*, *Cp*, and ground-water evapotranspiration were calculated for each pixel, an area of 8,743 ft^2 or about 0.20 acre, of the Landsat scene inside the phreatophyte boundary area. The volume of ground-water evapotranspiration and the area for each plant-cover zone were determined by summing the results within each zone.

Typical results are shown in figure B3 and given in table B1 for the Upper Fish Lake quadrangle in Little Fish Lake Valley (fig. B1). About 2,600 acres of the quadrangle are covered by phreatophyte shrubs and grasses. Phreatophyte plant cover in the quadrangle in 1985 ranged from an average of about 8 percent ($Cp = 0.08$) in the less than 10 percent zone to an average of about 60 percent ($Cp = 0.60$) in the greater than 50 percent zone, although coverage for individual pixels ranged from no plant cover (bare soil) to as much as 75 percent ($Cp = 0.75$) cover. Estimated ground-water evapotranspiration by phreatophytes was about 4,000 acre-ft.

The total water area of Little Fish Lake and Upper Fish Lake in Fish Lake Valley was about 800 acres in June 1985 (Chapter C of this report). Of this, 537 acres of surface water were within the limits of the Upper Fish Lake quadrangle—462 acres of Upper Fish Lake and 75 acres of Little Fish Lake. Upper Fish Lake commonly contains no water, but 1985 followed several years of above-normal precipitation. Little Fish Lake is a perennial lake, one of the few in the area, but most of the lake is south of the Upper Fish Lake quadrangle. Its total size ranged from 269 acres in 1985 to 187 acres in 1989 (Chapter C). Open-water evaporation in this area of Nevada may be as much as 4.5 ft/yr (Scott, 1971), but for Little Fish Lake it is not known how much of this represents ground water. It has been estimated that about 500 acre-ft/yr of ground water is evaporated from all of Little Fish Lake (Chapter C). Perhaps as much as 100 acre-ft of ground water was evaporated in 1985 from that part of Little Fish Lake within the limits of the Upper Fish Lake quadrangle.

Few ways are available by which to corroborate ground-water evapotranspiration estimated by the method described in this report. One way is to use this method to estimate evapotranspiration from areas near large springs and compare the estimated evapotranspiration to measured spring discharge. Large regional springs exist at Duckwater (fig. B4) and near Lockes (fig. B5) in Railroad Valley. Discharge from these springs appear to support large areas of phreatophyte shrubs and grasses. All or most of the measured discharge from these springs is assumed to be consumed by evapotranspiration by these phreatophytes. Discharge from the springs was measured periodically from 1968 to 1972 and from 1982 to 1988.

The areas near Duckwater and Lockes in which plant cover is affected by measured discharge from the springs can be delineated reasonably well on the plant cover map (figs. B4 and B5) derived from the Landsat image on the basis of plant cover, topography, and spring location. Estimated ground-water evapotranspiration from the Duckwater area (fig. B4) and for the area estimated to represent evapotranspiration of measured spring discharge are given in table B2. Ground-water evapotranspiration from the area west of the spring discharge area near Duckwater (fig. B4) may be associated with shallow ground water related to the nearby regional springs, but would not be supported by any of the measured discharge from those springs and is not included in the area believed to be affected by measured spring discharge. Estimated ground-water evapotranspiration of about 11,500 acre-ft is assumed to represent spring discharge at Duckwater and is com-

Table B1. Estimated ground-water evapotranspiration from the Upper Fish Lake area, Little Fish Lake Valley, Nevada, 1985 [1]

[Symbol: --, no data; E, estimated.]

	Average plant cover	Area (acres)	Evapotranspiration			
			Annual average rate (feet)	Summer (acre-feet)	Winter (acre-feet)	Annual (acre-feet)
Open water	0	537	E			100 E
Playa/bare soil	0	26	0.150	--	--	4
Cover <10 percent	.084	319	.632	142	59	201
Cover 10 to <20 percent	.145	1,510	1.401	1,534	582	2,116
Cover 20 to <35 percent	.258	530	2.148	850	288	1,138
Cover 35 to <50 percent	.418	132	2.504	253	78	331
Cover >50 percent	.600	89	2.588	179	51	230
Total		3,143		2,958	1,058	[2]4,100

[1] Calculated values are not rounded, to minimize rounding errors in the subsequent calculations, except as indicated.

[2] Rounded to nearest 100 acre-feet.

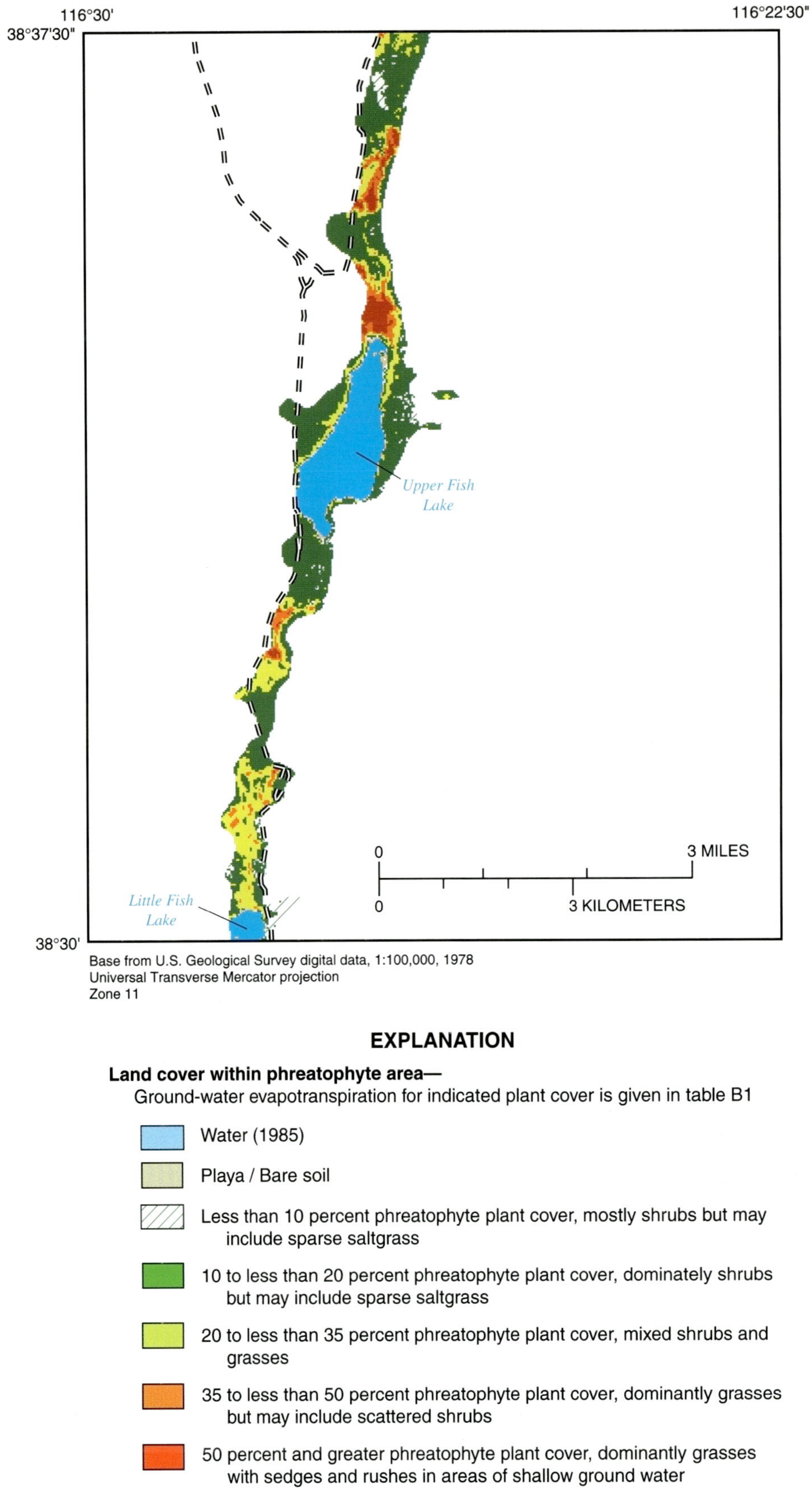

Figure B3. Estimated phreatophyte plant cover in Upper Fish Lake area, Little Fish Lake Valley, Nevada, 1985.

Table B2. Estimated ground-water evapotranspiration from the Duckwater Springs and Lockes area springs, Railroad Valley, Nevada, 1985 [1]

[Symbol: --, no data; E, estimated]

	Average plant cover	Area (acres)	Evapotranspiration			
			Annual average rate (feet)	Summer (acre-feet)	Winter (acre-feet)	Annual (acre-feet)
Railroad Valley - Duckwater area						
Open water		1	--	--	--	3 E
Playa/bare soil		40	0.150	--	--	6
Cover <10 percent	0.064	3,837	.383	1,032	436	1,468
Cover 10 to <20 percent	.139	1,346	1.336	1,301	497	1,798
Cover 20 to <35 percent	.277	1,004	2.215	1,667	557	2,224
Cover 35 to <50 percent	.424	927	2.511	1,781	545	2,326
Cover >50 percent	.706	1,785	2.580	3,609	995	4,604
Total		8,940		9,390	3,030	[2] 12,500
Railroad Valley - Duckwater Springs						
Open water		1	--	--	--	3 E
Playa/bare soil		23	.150	--	--	3
Cover <10 percent	.073	1,574	.497	551	231	782
Cover 10 to <20 percent	.141	1,176	1.350	1,149	438	1,587
Cover 20 to <35 percent	.277	994	2.216	1,651	552	2,203
Cover 35 to <50 percent	.424	923	2.511	1,774	543	2,317
Cover >50 percent	.707	1,779	2.580	3,598	992	4,590
Total		6,470		8,723	2,756	[2]11,500
Railroad Valley - Lockes area						
Open water		50	--	--	--	150 E
Playa/bare soil		21,602	.150	--	--	3,240
Cover <10 percent	.042	13,180	.189	1,748	748	2,496
Cover 10 to <20 percent	.136	1,668	1.303	1,571	603	2,174
Cover 20 to <35 percent	.259	394	2.152	634	215	849
Cover 35 to <50 percent	.398	70	2.483	132	41	173
Cover >50 percent	.543	12	2.582	25	7	32
Total		36,976		4,110	1,614	[2]9,100
Railroad Valley - Lockes area springs						
Open water		21	--	--	--	63 E
Playa/bare soil		315	.150	--	--	47
Cover <10 percent	.061	1,467	.380	392	165	557
Cover 10 to <20 percent	.140	758	1.344	738	281	1,019
Cover 20 to <35 percent	.262	302	2.164	488	165	653
Cover 35 to <50 percent	.397	62	2.482	117	36	153
Cover >50 percent	.545	10	2.583	20	6	26
Total		2,935		1,755	653	[2]2,500

[1] Calculated values are not rounded, to minimize rounding errors in subsequent calculations, except as indicated.

[2] Numbers are rounded to the nearest 100 acre-ft for values less than 10,000 acre-ft and to the nearest 500 acre-ft for values greater than 10,000 acre-ft.

pared with measured spring discharge of about 13,500 acre-ft in table B3. Estimated ground-water evapotranspiration is somewhat less than the measured spring discharge, but given the large area influenced by this spring discharge it is possible that the delineated area shown on figure B4 is in error or that some of the measured spring discharge infiltrates back into the shallow ground-water system and moves beyond the local phreatophyte area before it is evapotranspired by phreatophytes within the phyreatophyte boundary. Additionally, the measurements of spring discharge may be in error by 5 to 15 percent. The reason for the discrepancy is probably a combination of all these sources of error.

Estimated ground-water evapotranspiration from the Lockes area and from the area influenced by measured spring discharge near Lockes (fig. B5) also are given in table B2. Substantial areas of phreatophytes occur in the Lockes area (fig. B5) and are probably supported by the shallow ground water associated with the springs in the area. The area believed to be influenced by measured discharge from the springs near Lockes is shown in figure B5. The estimated ground-water evapotranspiration of about 2,500 acre-ft from this area is within the range of spring discharge of about 2,400 acre-ft measured in 1985 and 1986. The same possible sources of error in the estimated evapotranspiration and measured spring discharge described for the Duckwater area apply to the estimates and measurements for the Lockes area as well.

The good comparison between measured spring discharge and estimated ground-water evapotranspiration from the spring discharge areas at Duckwater and near Lockes suggests that the estimates of regional ground-water evapotranspiration, which are based on plant cover derived from Landsat *MSAVI* data, are reasonable.

SUMMARY AND CONCLUSIONS

The method of estimating ground-water discharge by evapotranspiration from phreatophytes for ground-water studies conducted in Nevada from the mid-1940's to the present used the results of field studies published in 1912 and 1932. Ground-water budgets based on these discharge estimates are used to allocate ground-water resources in Nevada. The results of more recent studies of evapotranspiration by rangeland phreatophytes using micrometeorological methods provide new data on which to base estimates of ground-water evapotranspiration.

An approach correlating ground-water evapotranspiration with plant cover was used in conjunction with a modified soil-adjusted vegetation index derived from Landsat data to develop a method for estimating the magnitude and distribution of ground-water evapotranspiration at a regional scale. Ground-water evapotranspiration was estimated using this method for two areas in Railroad Valley in which all or most of measured spring discharge is assumed to be consumed by evapotranspiration. Ground-water evapotranspiration of about 11,500 acre-ft/yr estimated by the method described herein compares well with measured discharge of about 13,500 acre-ft/yr from regional springs at Duckwater. Measured discharge from springs near

Table B3. Comparison of measured spring discharge and estimated evapotranspiration of spring discharge, Railroad Valley, Nevada, 1985 [1]

[Values of discharge and evapotranspiration in acre-feet per year]

Spring location	Date of measurement	Spring discharge	Date of measurement	Spring discharge	Estimated evapotranspiration, this study
Duckwater [2]	1-19-1985	14,000	2-3-1986	13,000	11,500
Lockes [3]	1-20-1985	2,500	2-2-1986	2,300	2,500

[1] Numbers are rounded to the nearest 100 acre-ft for values less than 10,000 acre-ft and to the nearest 500 acre-ft for values greater than 10,000 acre-ft.

[2] Includes discharge from Big Warm Spring and Little Warm Spring (Savard and Crompton, 1993, p. 78).

[3] Includes discharge from Hay Corral, North, Big, and Reynolds springs (Savard and Crompton, 1993, p. 77-78).

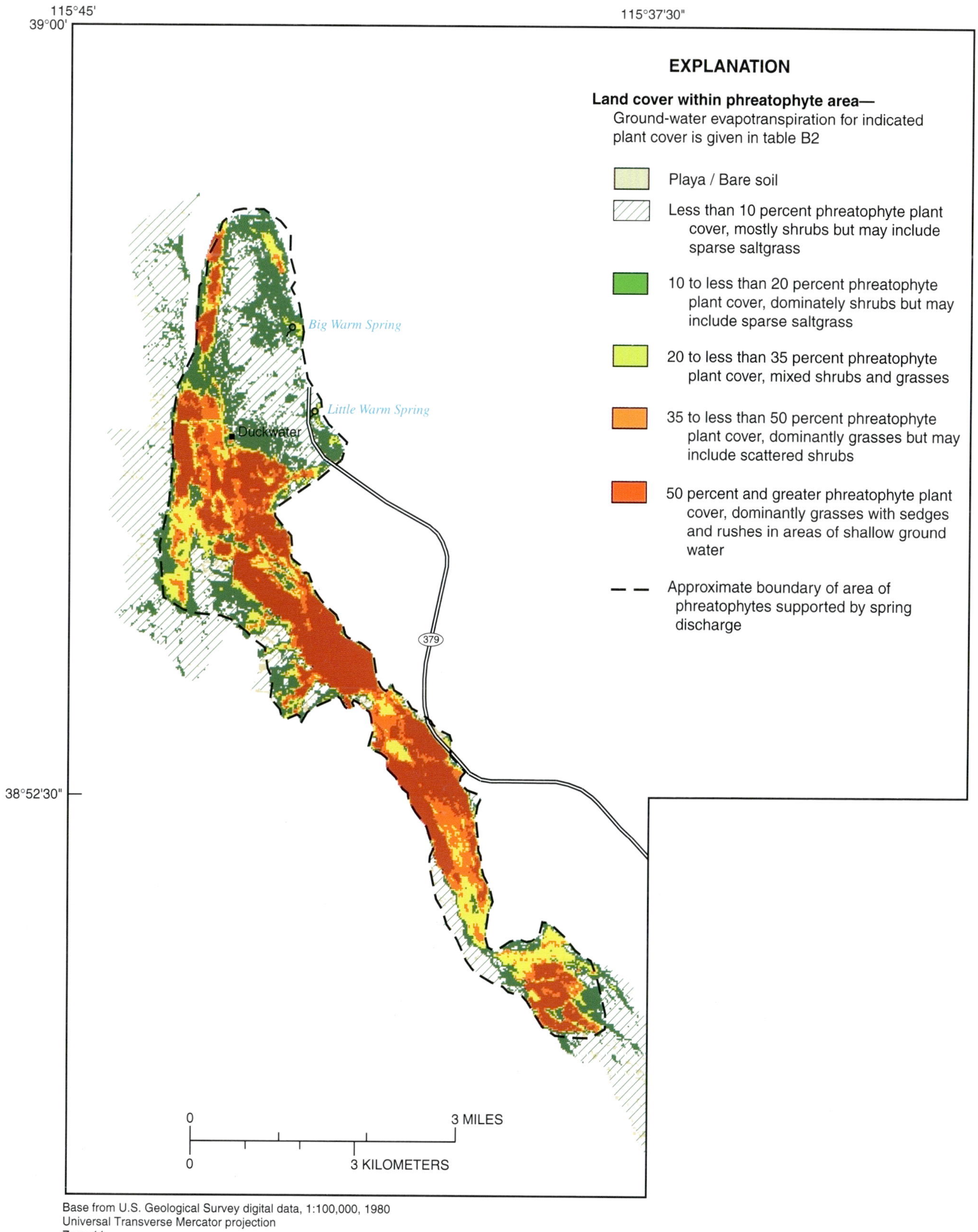

Figure B4. Estimated phreatophyte plant cover in the Duckwater area, Railroad Valley, Nevada, 1985.

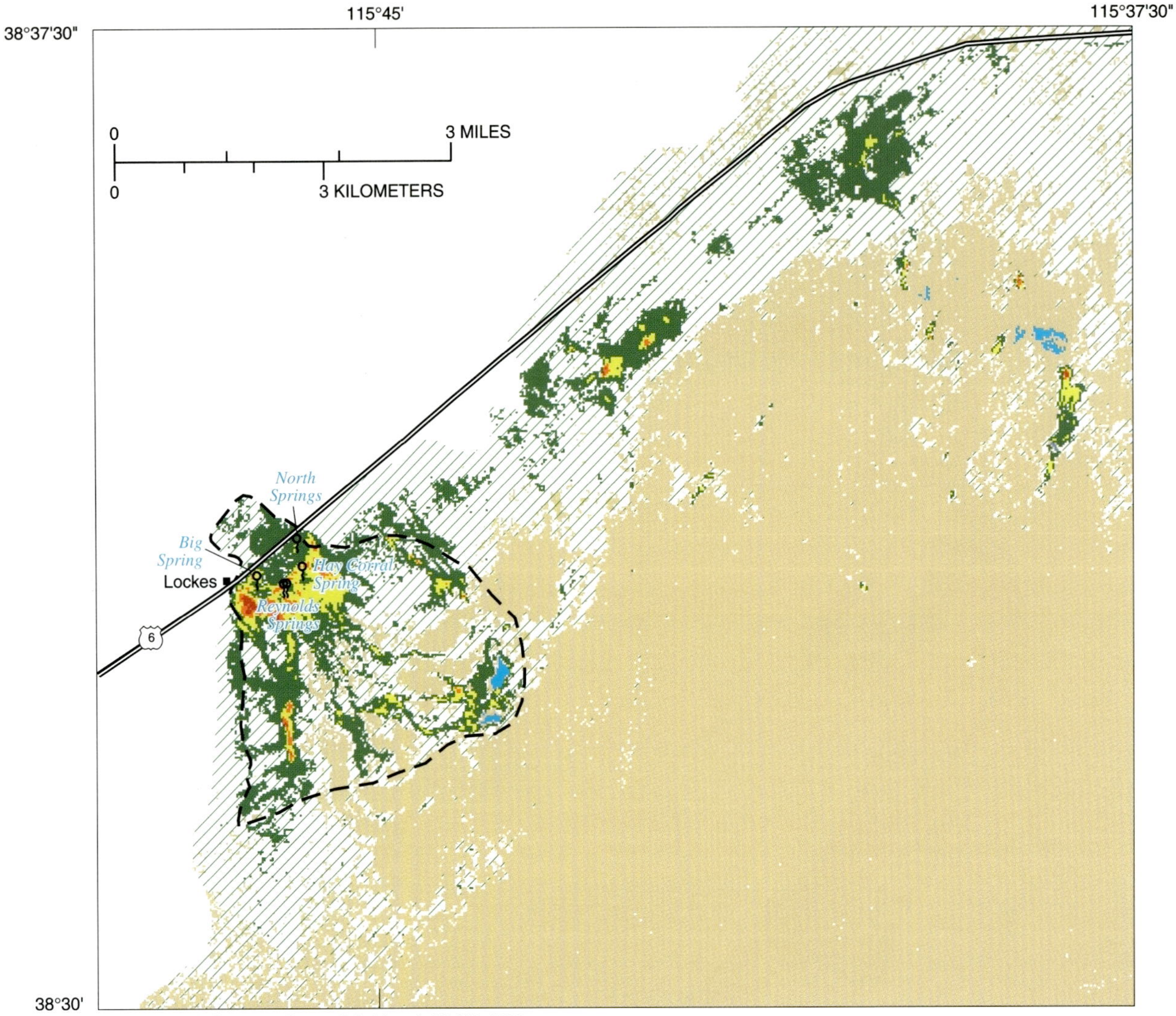

Base from U.S. Geological Survey digital data, 1:100,000, 1980
Universal Transverse Mercator projection
Zone 11

EXPLANATION

Land cover within phreatophyte area—Ground-water evapotranspiration for indicated plant cover is given in table B2

- Water
- Playa / Bare soil
- Less than 10 percent phreatophyte plant cover, mostly shrubs but may include sparse saltgrass
- 10 to less than 20 percent phreatophyte plant cover, dominately shrubs but may include sparse saltgrass
- 20 to less than 35 percent phreatophyte plant cover, mixed shrubs and grasses
- 35 to less than 50 percent phreatophyte plant cover, dominantly grasses but may include scattered shrubs
- 50 percent and greater phreatophyte plant cover, dominantly grasses with sedges and rushes in areas of shallow ground water
- — — Approximate boundary of area of phreatophytes supported by spring discharge

Figure B5. Estimated phreatophyte plant cover in the Lockes area, Railroad Valley, Nevada, 1985.

Lockes was about 2,400 acre-ft/yr; estimated groundwater evapotranspiration using the proposed method was about 2,500 acre-ft/yr.

REFERENCES CITED

Carman, R.L., 1989, Measurement of evapotranspiration in phreatophyte areas, Smith Creek Valley and Carson Desert, west-central Nevada, 1983: U.S. Geological Survey Water-Resources Investigation Report 89-4118, 18 p.

Chavez, P.S., 1989, Radiometric calibration of Landsat Thematic Mapper multispectral images: Photogrammetric Engineering and Remote Sensing, v. 55, no. 9, p. 1285-1294.

Duell, L.F.W., Jr., 1990, Estimates of evapotranspiration in alkaline scrub and meadow communities of Owens Valley, California, using the Bowen-ratio, eddy-correlation, and Penman-combination methods: U.S. Geological Survey Water-Supply Paper 2370-E, 39 p.

Groeneveld, D.P., and Warren, D.C., 1992, Total transpiration from land areas estimated from hand-held porometer measurements, Chapter E, *in* Wilson, D.H., Reginato, R.J., and Hollett, K.J., eds., Evapotranspiration measurements of native vegetation, Owens Valley, California, June 1986: U.S. Geological Survey Water-Resources Investigations Report 91-4159, p. 49-59.

Harrill, J.R., Gates, J.S., and Thomas, J.M., 1988, Major ground-water flow systems in the Great Basin region of Nevada, Utah, and adjacent states: U.S. Geological Survey Hydrologic Investigations Atlas HA-694-C, 2 sheets.

Huete, A.R., 1988, A soil-adjusted vegetation index (SAVI): Remote Sensing of Environment, v. 25, p 295-308.

Lee, C.H., 1912, An intensive study of the water resources of a part of Owens Valley, California: U.S. Geological Survey Water-Supply Paper 294, 135 p.

Lillesand, T.M., and Kiefer, R.W., 1987, Remote sensing and image interpretation: New York, John Wiley, 721 p.

Malek, E., Bingham, G.B., and McCurdy, G.D., 1990, Evapotranspiration from the margin and moist playa of a closed desert valley: Journal of Hydrology, v. 120, p. 15-34.

Monteith, J.L., and Unsworth, M.H., 1990, Principles of environmental physics (2d ed.): New York, Edward Arnold, 291 p.

Nichols, W.D., 1992, The uncertainty of water budget estimates in the Great Basin, *in* Hermann, R., ed., Managing water resources during global change: American Water Resources Association 28th Annual Conference and Symposium, Reno, Nev., Nov. 1-5, 1992, p. 309-317.

———1993, Estimating discharge of shallow groundwater by transpiration from greasewood in the northern Great Basin: Water Resources Research, v. 29, no. 8, p. 2771-2778.

———1994, Groundwater discharge by phreatophyte shrubs in the Great Basin as related to the depth to groundwater: Water Resources Research, v. 30, no. 12, p. 3265-3274.

Nichols, W.D., Laczniak, R.J., DeMeo, G.A., and Rapp, T.R., 1997, Estimated ground-water discharge by evapotranspiration, Ash Meadows area, Nye County, Nevada, 1994: U.S. Geological Survey Water-Resources Investigations Report 97-4025, 13 p.

Nichols, W.D., and Rapp, T.J., 1996, Micrometeorological data for energy budget and climate studies near Rogers Spring, Ash Meadows National Wildlife Refuge, Nye County, Nevada: U.S. Geological Survey Open-File Report 96-170, 49 p.

Qi, J., Chehbouni, A., Huete, A.R., Kerr, Y.H., and Sorooshian, S., 1994, A modified soil adjusted vegetation index: Remote Sensing of Environment, v. 48, p. 119-126.

Robinson, T.W., 1970, Evapotranspiration by woody phreatophytes in the Humboldt River valley near Winnemucca, Nevada: U.S. Geological Survey Professional Paper 491-D, 41 p.

Rush, F.E., 1968, Index of hydrographic areas: Nevada Division of Water Resources, Information Series Report 6, 38 p.

Sabins, F.F., Jr., 1987, Remote sensing principles and interpretation (2d ed.): New York, W.H. Freeman and Co., 449 p.

Savard, C.S., and Crompton, E.J., 1993, Hydrologic data for east-central Nevada, water years 1982-88: U.S. Geological Survey Open-File Report 90-153, 128 p.

Scott, B.R., 1971, Average annual evaporation: Nevada Division of Water Resources Map, scale 1:2,000,000.

White, W.N., 1932, A method of estimating ground-water supplies based on discharge by plants and evaporation from soil – results of investigations in Escalante Valley, Utah, *in* Contributions to the hydrology of the United States 1932: U.S. Geological Survey Water-Supply Paper 659, p. 1-105.

Chapter C. Regional Ground-Water Budgets and Ground-Water Flow, Eastern Nevada

By William D. Nichols

Title-page photograph. Quinn Canyon Range (right two-thirds of photo) and Southern Grant Range (left one-third of photo), Railroad Valley, Nevada, June 1994. View south from about 3 miles south of Lockes. Cliff-forming limestones of Cambrian Windfall Formation and Ordovician Lower Pogonip Group are exposed near summit of high peaks of northern Quinn Canyon Range (left of center) and near top of more distant ridge of southern Grant Range (left center). Scattered vegetation in foreground is phreatophytic greasewood. Valley-wide evapotranspiration of ground water may total about 85,000 acre-feet per year for Railroad Valley, of which almost 30 percent may be supplied by interbasin ground-water flow from adjacent valleys. Photograph by William D. Nichols, U.S. Geological Survey.

CONTENTS

PLATES

[Plates in pocket at back of report.]

FIGURES

TABLES

Chapter C. Regional Ground-Water Budgets and Ground-Water Flow, Eastern Nevada

By William D. Nichols

ABSTRACT

Previous estimates of ground-water budgets in Nevada were based on methods and data that now are more than 60 years old. Newer methods, data, and technologies were used in the present study to estimate ground-water recharge from precipitation and ground-water discharge by evapotranspiration by phreatophytes for 16 contiguous valleys in eastern Nevada. Annual ground-water recharge to these valleys was estimated to be about 855,000 acre-feet and annual ground-water evapotranspiration was estimated to be about 790,000 acre-feet; both are a little more than two times greater than previous estimates. The imbalance of recharge over evapotranspiration represents recharge that either (1) leaves the area as interbasin flow or (2) is derived from precipitation that falls on terrain within the topographic boundary of the study area but contributes to discharge from hydrologic systems that lie outside these topographic limits.

A vegetation index derived from Landsat-satellite data was used to estimate phreatophyte plant cover on the floors of the 16 valleys. The estimated phreatophyte plant cover then was used to estimate annual ground-water evapotranspiration. Detailed estimates of summer, winter, and annual ground-water evapotranspiration for areas with different ranges of phreatophyte plant cover were prepared for each valley. The estimated ground-water discharge from 15 valleys, combined with independent estimates of interbasin ground-water flow into or from a valley, were used to calculate the percentage of recharge derived from precipitation within the topographic boundary of each valley. These percentages then were used to estimate ground-water recharge from precipitation within each valley.

Ground-water budgets for all 16 valleys were based on the estimated recharge from precipitation and estimated evapotranspiration. Any imbalance between estimated recharge and estimated discharge may arise from errors in estimated ground-water evapotranspiration, from errors in the estimated precipitation in the topographic basin, or, more likely from a combination of errors in the two estimates. Imbalance between recharge and discharge in any valley was corrected by assuming the difference was equal to interbasin ground-water flow into or out of the valley. The proposed interbasin flow includes all errors in estimated ground-water evapotranspiration and recharge. Estimates of interbasin flow were available from previous studies for most valleys. For some valleys, the present study is the first to propose the interbasin flow or to propose an increase in the quantity of interbasin flow. Proposed areas of interbasin flow are consistent with available geologic, geophysical, and hydrologic evidence, but the suggested interbasin flow is not substantiated by such evidence.

INTRODUCTION

Ground-water resources in Nevada are managed on the basis of estimated ground-water budgets. The volume of ground water that may be allocated for development is assumed to equal an estimate of long-term mean annual ground-water recharge to a given hydrographic area or basin (Rush, 1968). Ground-water recharge to the topographically closed basins and other hydrographic areas of Nevada is from precipitation and may be augmented by interbasin flow or surface-water inflow from adjacent basins. Measuring ground-water recharge is difficult at local scales and not possible at regional scales with current technology.

Consequently, rather than measuring recharge, the approach in Nevada has been to estimate it on the basis of ground-water discharge from a basin or hydrographic area. Before the development of ground water began in any given valley, the ground-water system was assumed to have been in equilibrium—long-term mean annual ground-water discharge was assumed to be equal to the long-term mean annual recharge. Ground-water discharge from a valley or hydrographic area is by evapotranspiration from phreatophytes and bare soil or by interbasin flow to adjacent valleys. Interbasin ground-water flow to and from a basin may be surmised from water-level gradients, estimates of hydraulic conductivity, and estimates of aquifer thickness or from estimated ground-water budget imbalances. Ground-water evapotranspiration can be measured at the local scale using micrometeorological methods.

Estimates of ground-water evapotranspiration and estimates of interbasin ground-water flow from a valley have been the foundations for estimating regional ground-water recharge as a function of annual precipitation. Reconnaissance-level studies of the ground-water resources of Nevada began in 1945 and continued into the 1970's. These studies estimated ground-water evapotranspiration by mapping the areas covered by phreatophytes and applying evapotranspiration rates from early investigations of ground-water evapotranspiration from saltgrass by Lee (1912) in Owens Valley, Calif., and ground-water evapotranspiration from phreatophyte shrubs by White (1932) in Escalante Desert, Utah. Many of the valleys, or groups of adjacent valleys, studied during the late 1940's when the reconnaissance methods for estimating ground-water budgets were being developed, were assumed to be hydrologically closed; interbasin flow into or out of the valleys was assumed not to occur and was thus not included in the ground-water budgets. Precipitation falling in a basin was estimated from a map of annual precipitation in Nevada (Hardman, 1936). Using estimates of annual ground-water discharge from several valleys together with estimates of annual precipitation in the same valleys, a method was devised for estimating annual ground-water recharge as a function of annual precipitation (Maxey and Eakin, 1949; Eakin and others, 1951b, p. 26 and 80).

A recently developed method for estimating ground-water evapotranspiration at regional scales (Chapter B of this report) and a recent map of annual precipitation in Nevada (G.H. Taylor, Oregon Climate Center, written commun., May 1997) provide an opportunity to reassess the ground-water budgets estimated by previous studies in Nevada.

Values of estimated ground-water evapotranspiration and estimated recharge from precipitation given in tables C5, C7, and C13 have been rounded using the following rules. Numbers are rounded to the nearest 100 acre-ft for values less than 10,000 acre-ft, to the nearest 500 acre-ft for values equal to or greater than 10,000 acre-ft and less than 100,000 acre-ft, and to the nearest 1,000 acre-ft for values equal to or greater than 100,000 acre-ft. Calculated values of ground-water evapotranspiration, volumes of precipitation, and recharge from precipitation given in tables C6, C9, C10, C11, C17, C18, and C19 are not rounded, to minimize rounding errors in subsequent calculations, to maintain precision, and do not imply accuracy.

Purpose and Scope

This chapter provides estimates of mean annual ground-water evapotranspiration at a regional scale, mean annual ground-water recharge from precipitation, and regional ground-water budgets and ground-water flow for selected valleys in eastern Nevada. The budgets were determined from estimates of ground-water evapotranspiration from phreatophyte plant cover using a vegetation index derived from Landsat data. Estimates of ground-water evapotranspiration combined with estimates of interbasin ground-water flow were used to estimate ground-water recharge as a function of precipitation. Ground-water budgets were estimated for 16 valleys in eastern Nevada and then used to define regional ground-water flow in that area. This study was conducted in cooperation with the Las Vegas Valley Water District and the Nevada Division of Water Resources.

Study Area

The study area covers 14,986 mi^2 of eastern Nevada (fig. C1), an area about 500 mi^2 larger than the area covered by the states of Massachusetts, Connecticut, and Rhode Island. The study area includes, from northwest to southeast, Ruby, Clover, Independence, Goshute, Antelope, Newark, Long, Butte, Steptoe, Tippett, Little Fish Lake, Hot Creek, Little Smoky, Railroad (northern part), Jakes, and Spring Valleys in Elko, White Pine, Eureka, Lincoln, and Nye Counties

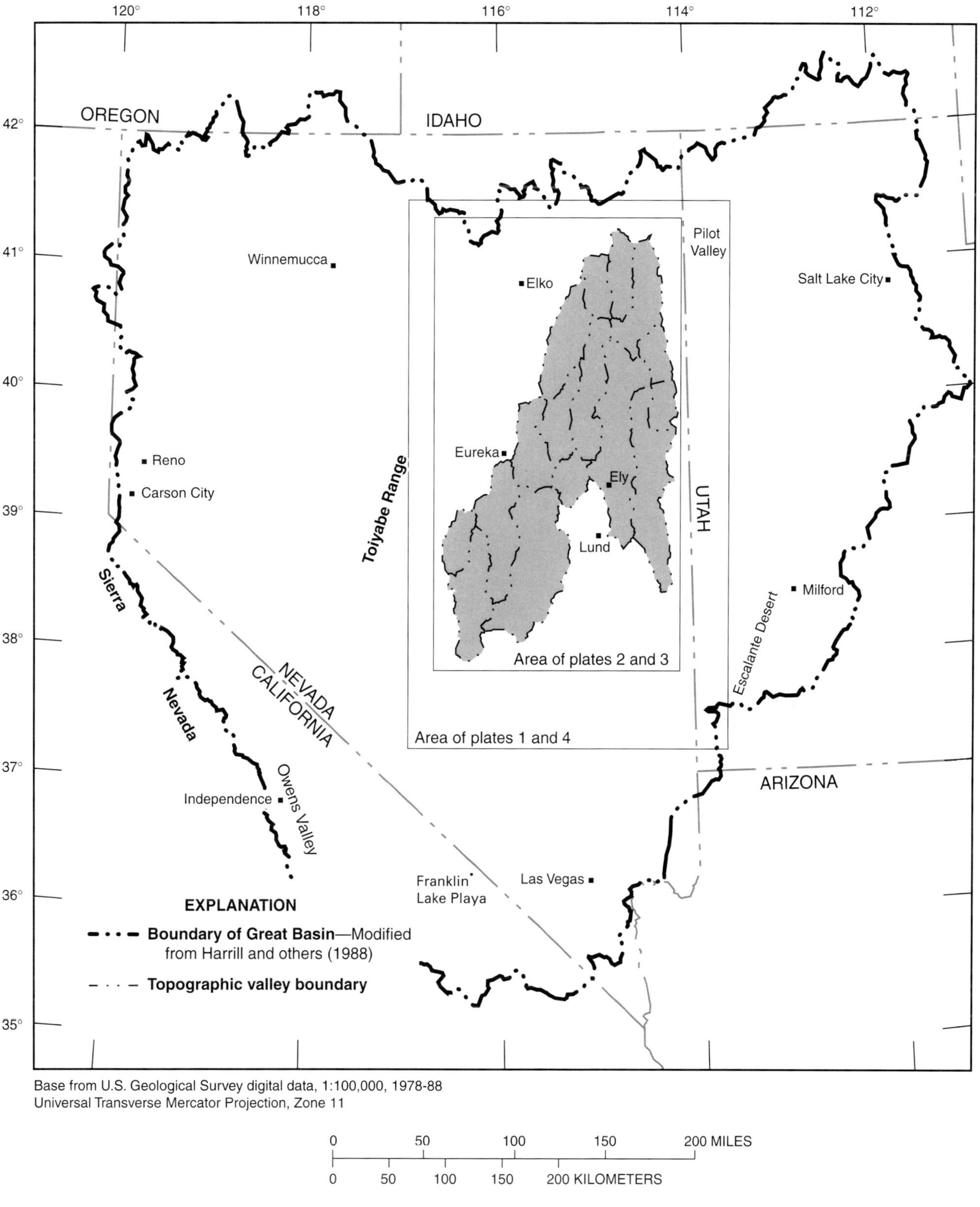

Figure C1. Great Basin and eastern Nevada study area.

(pl. 1, table C1). The valleys approximately correspond to hydrographic areas[1] of the same name as defined by Rush (1968).

Geography and Topography

The study area covers a sparsely populated region of the central Great Basin in eastern Nevada (fig. C1, pl. 1). Several valleys have no permanent residents and others have only a few isolated ranches. Small areas of irrigated agriculture in Ruby, Clover, Newark, Steptoe, Spring, Little Smoky, and Railroad Valleys use ground water that either is pumped or is discharged by springs. Ruby and Clover Valleys have extensive areas of subirrigated native pasture along the northwest part of each valley; most of this water is derived from surface runoff from the northern Ruby Mountains and East Humboldt Range. Much of the remaining rangeland in the study area is used for cattle grazing.

Mining has been ubiquitous throughout the area; 52 mining districts are within the study area (Smith, 1976; Kleinhampl and Ziony, 1984; LaPointe and others, 1991). Mining activities began in the 1860's and continue to the present (1999) at several locations. Elements recovered, in order of occurrence, but not in value, are lead, silver, tungsten, copper, and gold.

The city of Ely (pop. 4,756 in 1990) is in the southern part of the study area in Steptoe Valley (fig. C1; pl. 1). The towns of McGill (pop. 1,258 in 1990), Ruth (pop. unknown), and Duckwater (pop. 298 in 1990) are the only other populated communities in the study area (pl. 1). Eureka (pop. 1,107 in 1990) is just west of the study area; Elko (pop. 14,736 in 1990) and Wells (pop. 1,256 in 1990) are to the northwest and north, respectively (pl. 1). Interstate 80 crosses the northern end of Independence and Goshute Valleys; U.S. Highway 50 crosses the central part of the study area from Eureka to Ely (pl. 1) and thence eastward into Utah. U.S. Highway 6 crosses the southern part of the study area from Hot Creek Valley through Railroad Valley to Ely in Steptoe Valley. U.S. Highway 93 connects Ely northward to Wells and Interstate 80, and southward toward southern Nevada.

[1]Formal hydrographic areas in Nevada were delineated systematically by the U.S. Geological Survey and Nevada Division of Water Resources in the late 1960's (Rush, 1968; Cardinalli and others, 1968) for scientific and administrative purposes. The official hydrographic-area names, numbers, and geographic boundaries continue to be used in Geological Survey scientific reports and Division of Water Resources administrative activities.

Valleys and bounding mountain ranges of the study area trend generally north-south with the exception of the Ruby Mountains and East Humboldt Ranges on the north and the Quinn Canyon and Grant Ranges on the south, which trend northeast-southwest. The defining features of the central Great Basin in eastern Nevada are valleys whose lowest areas are nearly a mile or more above sea level with bounding mountain ranges that rise as much as a mile or more above the adjacent valley. The altitude of the valley floors (table C1) ranges from 4,700 ft (Railroad Valley) to 6,500 ft (Little Fish Lake Valley) above sea level and averages about 5,800 ft above sea level. Mountain ranges bounding the valleys of the study area rise to altitudes of more than 10,000 ft in many locations. They include the rather low Pancake Range (pl. 1), which varies from 7,000 to 8,000 ft, and rises to 9,240 ft at Portuguese Mountain, to the Ruby Mountains, which have large areas above 10,000 ft and rise to over 11,000 ft at King Peak (pl. 1), to the Snake Range, which rises to as much as 13,063 ft at Wheeler Peak (pl. 1). Clover, Independence, Newark, Railroad, Ruby, Spring, and Steptoe Valleys have large, well-defined playas. Little Fish

Table C1. Names and areas of valleys included in eastern Nevada study area [1]

Valley name	Area: Acres	Area: Square miles	Altitude of lowest area of valley floor (feet above sea level)
Antelope Valley	255,683	399.5	5,600
Butte Valley	652,333	1,019.3	6,000
Clover Valley	292,099	456.4	5,600
Goshute Valley	612,146	956.5	5,600
Hot Creek Valley	658,493	1,028.9	5,200
Independence Valley	360,677	563.6	5,600
Jakes Valley	270,512	422.7	6,300
Little Fish Lake Valley	276,471	432.0	6,500
Little Smoky Valley	740,552	1,157.1	6,000
Long Valley	419,850	656.0	6,100
Newark Valley	509,264	795.7	5,900
Railroad Valley (northern part)	1,369,718	2,140.2	4,700
Ruby Valley	638,923	998.3	6,000
Spring Valley	1,066,995	1,667.2	5,600
Steptoe Valley	1,245,626	1,946.3	5,900
Tippett Valley	221,588	346.2	5,700
Total [2]	9,590,900	15,000	

[1] Calculated values are not rounded, to minimize rounding errors in subsequent calculations, except as indicated.

[2] Rounded to the nearest 100 acres or square miles.

Lake in Little Fish Lake Valley and Ruby Lake in Ruby Valley are the only natural permanent lakes in the study area. Snow Water Lake in Clover Valley, Franklin Lake in northern Ruby Valley, Goshute Lake in Steptoe Valley, and Newark Lake in Newark Valley are present in many years, but disappear during periods of extended drought. Long and Tippett Valleys each have a small playa underlain by deep ground water. Many playas have water on them during the spring and early summer, especially in years of above average precipitation. Some playas become intermittent lakes during extended wet periods.

Hydrogeology

The geology of eastern Nevada is complex and has involved repeated episodes of sedimentation, folding, faulting, and volcanic activity. Consolidated rocks in the study area, ranging in age from Precambrian to late Tertiary, are exposed in the mountain ranges bounding the valleys. Unconsolidated deposits of Tertiary and Quaternary age underlie the valleys of the study area. The geologic framework of the region as it relates to ground-water hydrology has been summarized by Plume and Carlton (1988) and Plume (1996).

Precambrian rocks, dominantly quartzite, occur in small areas of the southern Cherry Creek Range, northern Egan Range, and in large areas of the Schell Creek and Snake Ranges (Hose and Blake, 1976). Lower Paleozoic (Cambrian to Devonian) limestone and dolomite with minor shale, siltstone, sandstone, and quartzite comprise major parts of the southern Ruby Mountains, the Goshute Mountains, the southern Cherry Creek and the northern and southern Egan Ranges, the central Schell Creek Range, and the Fish Creek, Snake, White Pine, and Grant Ranges. Upper Paleozoic (Pennsylvanian to Permian) limestone and cherty limestone with interbedded shale and locally thick interbedded sandstone comprise major parts of the Pequop, Diamond, Maverick Springs, Butte, and Spruce Mountains, and the northern Cherry Creek, central Egan, and southern Schell Creek Ranges. The East Humboldt Range and the northern half of the Ruby Mountains consist of a complex of metamorphosed Paleozoic rocks and Mesozoic granitic rocks. Intrusive igneous rocks of Mesozoic and Cenozoic age crop out in many of the mountain ranges of the study area. Tertiary volcanic rocks dominate the Monitor, Hot Creek, Pancake, Quinn Canyon, and both Antelope Ranges, and occur locally in a number of the other mountain ranges in the study area.

Winograd and Thordarson (1975) defined a hydrogeologic framework for southern Nevada (fig. C2). They combined the carbonate rocks of Middle Cambrian through Devonian age, which includes the upper Carrara Formation through the Devils Gate limestone, into a single hydrogeologic unit that they referred to as the lower carbonate aquifer. Overlying this aquifer, they defined an upper clastic aquitard that consists of the Devonian-Mississippian Eleana Formation. Above this aquitard, they defined an upper carbonate aquifer consisting of the Pennsylvanian-Permian Tippipah Limestone.

These same geologic formations and hydrogeologic units, or their stratigraphic and commonly lithologic equivalents, are found throughout the study area in eastern Nevada (fig. C2). Plume and Carlton (1988), and later Dettinger (1989), Plume (1996), and Dettinger and Schaefer (1996) concluded that the two carbonate aquifer units of Winograd and Thordarson (1975) commonly are in fault contact and probably are connected hydraulically throughout most of Nevada. Consequently, Plume and Carlton (1988) considered there is a single carbonate-rock aquifer that can be divided into a lower part, equivalent to the lower carbonate aquifer of Winograd and Thordarson (1975), and an upper part, equivalent to the upper carbonate aquifer of Winograd and Thordarson (1975). The hydrogeologic terminology of Winograd and Thordarson (1975) is used in this report.

The water-bearing properties of the Tertiary and Quaternary volcanic rocks overlying the carbonate-rock aquifers are not well known except in a few local areas (Winograd and Thordarson, 1975; Glancy, 1986). They are considered as a single hydrogeologic unit for this study. Unconsolidated Tertiary and Quaternary basin-fill deposits are the units that yield most of the ground water in the study area. The hydraulic properties of these deposits have been described in some detail by Plume (1996).

Climate

The climate of the valleys of eastern Nevada, as most of the Great Basin, is a middle-latitude desert and steppe climate dominated by continental tropical air masses in the summer and continental polar air masses in the winter (Houghton and others, 1975, p. 13, 69-70). The arid climate is characterized by hot summers and cold winters where annual precipitation commonly is less than 12 inches (table C2). The bounding mountain ranges receive more precipitation and have a

System	Series	Southern Nevada[1]	Southern Nevada hydrogeologic units[1]	Grant, Quinn Canyon, and Egan Ranges[2]	Snake Range[3]	Ruby Mountains[4]	Goshute Mountains[5]	Eastern Nevada hydrogeologic units[6]
Permian		Tippipah Limestone	Upper carbonate aquifer	Arcturus Formation	Arcturus Formation	Pequop Formation	Park City Group, undivided Pequop Formation	Upper part of carbonate-rock aquifer
Pennsylvanian				Riepe Spring Limestone	Riepe Spring Limestone	Limestone and dolomite	Riepe Spring Limestone	
				Ely Limestone	Ely Limestone	Ely Limestone	Ely Limestone	
Mississippian		Eleana Formation	Upper clastic aquitard	Diamond Peak Formation	Scotty Wash Quartzite	Diamond Peak Formation	Diamond Peak Formation	Clastic rocks
				Chainman Shale	Chainman Shale	Chainman Shale	Chainman Shale	
				Joana Limestone	Joana Limestone	Joana Limestone	Joana Limestone	
Devonian				Pilot Shale	Pilot Shale	Pilot Shale	(hiatus)	
		Devils Gate Limestone	Lower carbonate aquifer	Guilmette Formation	Guilmette Formation	Devils Gate Limestone	Guilmette Formation	Lower part of carbonate-rock aquifer
		Nevada Formation		Simonson Dolomite	Simonson Dolomite	Nevada Formation	Simonson Dolomite	
				Sevy Dolomite	Sevy Dolomite		Sevy Dolomite	
Devonian and Silurian				Laketown Dolomite	Laketown Dolomite	Lone Mountain Dolomite	Laketown Dolomite	
Ordovician	Upper	Ely Springs Dolomite		Ely Springs Dolomite	Fish Haven Dolomite	(hiatus)	Fish Haven Dolomite	
	Middle	Eureka Quartzite		Eureka Quartzite	Eureka Quartzite	Eureka Quartzite	Eureka Quartzite	
	Lower	Pogonip Group, undivided		Pogonip Group, undivided	Pogonip Group, undivided	Pogonip Group, undivided	Pogonip Group, undivided	
Cambrian	Upper	Nopah Formation		Windfall Formation (?)	Notch Peak Formation	Windfall Formation	Notch Peak Formation	
					Corset Spring Shale[8]		Windfall Formation (?)	
		Dunderberg Shale[7]		Dunderberg Shale[7]	Johns Wash Limestone[8]	Dunderberg Shale[7]	Dunderberg Shale[7]	
	Middle	Bonanza King Formation		Lincoln Peak Formation	Lincoln Peak Formation	Limestone, dolomite, and shale	Limestone and dolomite	
		Carrara Formation		Pole Canyon Limestone	Pole Canyon Limestone			
	Lower	Zabriskie Quartzite	Lower clastic aquitard	Pioche Shale	Pioche Shale	Pioche Shale	Pioche Shale	Clastic rocks

[1]Winograd and Thordarson (1975)
[2]Tschanz and Pampeyan (1970), Hose and Blake (1976), and Kleinhampl and Ziony (1984)
[3]Tschanz and Pampeyan (1970) and Hose and Blake (1976)
[4]Dudley (1967), Hose and Blake (1976), and Coats (1987)
[5]Coats (1987)
[6]Plume and Carlton (1988)
[7]Member of the Nopah Formation
[8]Member of the Orr Formation

Figure C2. Correlation of stratigraphic and hydrostratigraphic units in southern Nevada and the eastern Nevada study area.

Table C2. Climatic data for 1961-90 at selected valley locations in eastern Nevada. Locations are shown on plate 1

[Digital data from Western Regional Climate Center, Desert Research Institute, University of Nevada, Reno, 1997.]

Station name	Altitude (feet above sea level)	Mean annual temperature (degrees Fahrenheit)		Annual precip-itation (inches)
		Maximum	Minimum	
Arthur	6,300	57.6	30.4	15.33
Duckwater	5,400	--	--	7.38
Elko	5,075	62.4	30.9	9.95
Ely	6,262	61.2	28.0	10.13
Eureka	6,540	59.8	33.2	13.10
Lund	5,565	64.9	30.9	10.46
Ruby Lake	6,012	61.6	32.1	13.32
Snowball Ranch	7,160	59.0	28.9	9.01
Wells	5,650	59.6	28.5	10.59

subhumid continental climate characterized by cold winters with moderate annual precipitation ranging from 20 to 40 inches. The highest ranges in Nevada, including the Ruby Mountains and Snake Range within the study area, have a humid continental climate because of the abundant precipitation, greater than 30 inches, at the higher altitudes.

Summer maximum daily temperatures in the valleys of the study area typically exceed 90°F during late July and early August, and may exceed 100°F, with valleys at lower altitudes experiencing higher temperatures over longer periods of time. Daily minimum summer temperatures in most of the valleys range from the low 40's to low 50's, but in Steptoe Valley the summer minimum is about 10° lower. Maximum winter daily temperatures in the valleys range from 30° to 40°F in the northern part of the study area and are about 10° warmer in the southern part of the study area. Persistent temperature inversions develop in many of these valleys during some winters. Winter minimum daily temperatures commonly range from 10° to 20°F, but may go as low as –30° to –40°F especially in the northern part of the study area. Maximum and minimum daily temperatures decrease with altitude in the mountain ranges bounding the valleys. No temperature records are available for the high-altitude areas. However, in the absence of temperature inversions, the temperature lapse rate, which is the rate of temperature decrease with altitude, may be as much as 5.9°F per 1,000 feet on the basis of atmospheric soundings by the U.S. Weather Service at Winnemucca, Elko, and Ely, Nev. This suggests temperatures at higher altitude in the surrounding mountains may be 15° to 25°F cooler than in the valleys.

Vegetation

The diverse climate zones of the Great Basin and the study area lead to diverse vegetation zones. In the arid zones of the lower valley floors are plants of the salt desert and shadscale-greasewood communities. Plants of these two zones are discussed in more detail below.

Beyond the valley floors are alluvial fans leading to the base of the surrounding mountains. On the fans are plants of the Great Basin sagebrush community, which commonly includes big sagebrush (*Atremisia tridentata*), dwarf sagebrush (*A. arbuscula*), and black sagebrush (*A. arbuscula* var. *nova*), and may include spiny hopsage (*Grayia spinosa*), bitterbrush (*Purshia tridentata,* and *P. glandulosa*), and mormon tea (*Ephedra viridis*). Before the introduction of livestock, this zone included numerous species of grasses, with areas that received greater precipitation having a higher percentage of grass to shrub. Although grasses are common in this zone today, they may not be as extensive as they were originally (Trimble, 1989, p. 101, 110).

Beginning in the middle to upper part of the fans, between 6,000 and 7,000 ft in altitude, is the lower border of the pinyon-juniper woodland assemblage, which may extend to altitudes of nearly 9,000 ft (Charlet, 1996, p. 233). This assemblage is characterized in the Great Basin by generally open stands of singleleaf pinyon (*Pinus monophylla*) and Utah juniper (*Juniper oseosperma*) with an understory of sagebrush and cool-season grasses (Trimble, 1989). Juniper dominates in the lower altitudes near the valley floor with pinyon becoming more common with increasing altitude and generally dominating above 7,000-7,500 ft. Above about 7,500 ft, mountain mahogany (*Cercicarpus ledifolius* and *C. inticatus*) and aspen (*Populus tremuloides*) may be included in this assemblage.

A pine-fir forest should occur above the pinyon-juniper woodland, but many of the ranges in the central Great Basin, including some of those of the study area, have no forest above the woodland. Instead, many of the highest ranges, such as the Schell Creek and Snake Ranges, have a double timberline, with the lower timberline occurring between about 8,000 and 8,500 ft. Here, the pinyon-juniper woodland stops, but sagebrush, bitterbrush, and grasses together with small groves of curlleaf mountain mahogany and aspen fill the zone of the fir-pine forest for the next 500 to 1,500 ft in altitude to the lower border of the subalpine zone. The trees of the pine-fir forest zone are present in small stands on protected north-facing slopes and even at lower altitudes in well-watered washes and canyons.

The subalpine zone begins at about 9,000-9,500 ft and extends to the upper timberline between 11,000 and 11,500 ft. The trees of this zone include whitebark pine (*Pinus monticola*), limber pine (*Pinus flexilis*), bristlecone pine (*Pinus longaeva*), Engleman spruce (*Picea englemanii*), and aspen. Above timberline, alpine tundra is found in the highest ranges of the study area, including the Ruby Mountains, and the East Humboldt, Schell Creek, and Snake Ranges.

The plants of the salt desert and shadscale-greasewood communities found at the lowest altitudes in the valleys, however, are of most interest in the present study. These communities include the phreatophytes that consume and discharge ground water by evapotranspiration. Some valleys in the study area have a playa at the lowest part of the valley, which commonly is underlain by a shallow water table. The playas range in size from a small one in Long Valley (about 6,000 acres) to the large playa in Railroad Valley (about 36,000 acres). Surrounding these playas, or in the lowest parts of the valleys with no playa, are plants of the salt desert community, including iodine bush (also called pickleweed; *Allenrolfea occidentalis*), saltsage (*Atriplex tridentata*), and saltgrass (*Distichlis spicata* var. *stricta*). Iodine bush is reported to grow in areas with a depth to water of as much as 20 ft (Robinson, 1958, p. 32). Saltgrass, the most common phreatophyte in this zone, grows mostly in areas where the depth to water is less than about 8 ft, but has been reported to grow in areas where the water table is as much as 12 ft deep (Blaney and others, 1933, p. 50).

Beyond this fringe of vegetation, around the margin of the playas, is the shadscale-greasewood plant community. Within this plant association are found not only greasewood (*Sarcobatus vermiculatus*) and shadscale (*Atriplex confertifolia*), but also saltbush (also known as chamiso; *Atriplex canescens*), spiny hopsage (*Grayia spinosa*), winterfat (*Ceratoides lanata*), and where soils and ground water are less saline, rabbitbrush (*Chrysothamnus nauseosus*) and big sagebrush (*Artemisia tridentata*).

Shadscale and spiny hopsage are not commonly included as phreatophytes, but have been observed growing with greasewood in areas where the depth to water is at least 15 ft and therefore are assumed to transpire ground water at rates similar to those of greasewood. White (1932, p. 38) discusses the occurrence of shadscale with greasewood and rabbitbrush in areas of shallow ground water, but does not suggest a limiting depth to water. Saltbush is found where ground water is from about 8 ft to as much as 62 ft below land surface (Robinson, 1958, p. 33).

Greasewood is the principal phreatophyte, other than riparian, in the shadscale-greasewood zone of western Nevada (Billings, 1951) and its range is more extensive than that of big sagebrush in western North America (Robertson, 1983). Greasewood covers about 12 million acres from Canada to Mexico, but prefers the cold deserts north of 37°N latitude (Shreve, 1942). Greasewood occurs in areas where the depth to ground water ranges from about 5 ft to 35 ft and perhaps to as much as 60 ft (Robinson, 1958, p. 39). White (1932, p. 33) noted that greasewood required at least 3 ft of unsaturated soil most of the time.

Rabbitbrush grows in areas where the depth to ground water is less than about 35 ft. Robinson (1958, p. 34) suggests a maximum depth to water for rabbitbrush of 15 ft, conventional wisdom has suggested a maximum depth of 25 ft, and Mower and Nace (1957, p. 18) suggest a maximum depth of 35 ft.

Big sagebrush has been observed in some valleys of the Great Basin in association with rabbitbrush in areas where the water table is about 12 to 15 ft below land surface, and in these circumstances appears to be a phreatophyte as well. White (1932, p. 43), however, assumed on the basis of water-level fluctuations, that sagebrush used little or no ground water even in the Escalante Valley of Utah where the depth to water was less than 15 ft. Mozingo (1987, p. 271) reports that sagebrush has roots that grow as deep as 12 ft and commonly penetrate to the capillary zone above the water table.

Previous Ground-Water Studies

Studies to measure or estimate the consumptive use of ground water by phreatophytes, and thus a large part of the discharge component of the ground-water budget in the Great Basin, began in 1910 with the study by Lee (1912) and continued in the 1920's with the work of White (1932). Additional studies were done by other investigators through the 1930's (Blaney and others, 1930, 1938; summarized by Young and Blaney, 1942), 1940's (Gatewood and others, 1950), and continued into the 1960's with the study by Robinson

(1970). In the late 1940's, investigators in Nevada used the results of some of the earlier evapotranspiration studies to make estimates of the recharge component of the ground-water budget (Maxey and Eakin, 1949; Eakin and others, 1951b). These methods have been used in ground-water studies in Nevada up to the present time.

Ground-Water Discharge Studies

Among the earliest studies to measure the consumptive use of ground water by evapotranspiration from saltgrass and native shrubs and evaporation from bare soil were those of Lee (1912) and White (1932). Lee (1912) performed evapotranspiration-tank studies in 1910 and 1911 near Independence, Calif. (fig. C1), in the Owens Valley in an attempt to determine ground-water evapotranspiration from saltgrass and bare soil in relation to the depth to ground water. Similar evapotranspiration-tank studies were done by White (1932) in 1926 and 1927 to determine ground-water evapotranspiration from alfalfa, saltgrass, greasewood, and bare soil in the Escalante Desert near Milford, Utah (fig. C1).

These, as well as subsequent, evaporation-tank studies all used similar techniques, as discussed in Chapter A of this report (see section titled "Previous Studies"). Some of the studies were performed continuously for 12 months while others were operated only from April or May to October; the studies were from 1 or 2 years.

Lee's early study of saltgrass (1912) seemed to demonstrate a clear relation between evapotranspiration of ground water and the depth to ground water. Although subsequent studies (Blaney and others, 1930, 1938; Young and Blaney, 1942) produced similar relations, an analysis of these data by Weeks and others (1987) demonstrated that depth to ground water does not adequately define a unique relation to ground-water evapotranspiration by saltgrass. The locations of these evaporation-tank studies ranged from about 34° to about 38°N in latitude, from about 1,500 ft to over 3,000 ft in altitude, and from the coastal climate of southern California to the continental climate of southern Colorado. None of these studies provided a measure of the amount of vegetation, such as density or leaf area index.

Studies by White (1932) attempted to determine ground-water evapotranspiration from alfalfa, saltgrass, greasewood, and bare soil. White (1932) also installed a number of small-diameter monitoring wells in areas of native vegetation underlain by shallow ground water and measured small diurnal fluctuations of ground-water levels that were interpreted to be caused by the evapotranspiration of ground water during the day. An expression was developed relating the discharge of ground water to the specific yield of the water-bearing sediments and the 24-hour rate of water-level change plus net water-level change (White, 1932, p. 61). Soil samples were taken to determine appropriate values of specific yield. The volume of ground water discharged then was determined from the diurnal water-level fluctuations. This volume was related to biomass production and converted to areal estimates of ground-water discharge based, presumably, on areal estimates of plant canopy volume. White (1932, p. 86-87) concluded that, in the Escalante Desert, saltgrass and meadowgrass in areas where the depth to ground water was between 0 and 5 ft consumed about 1 ft of ground water a year, and that greasewood, rabbitbrush, and shadscale consumed about 0.15 ft/yr in areas with a depth to ground water of 8 to 30 ft, and about 0.42 ft/yr, when evaporation of ground water from bare soil was included, in areas with a depth to ground water of less than 8 ft. Evaporation of ground water from bare soil can be important in areas where the depth to ground water is as much as 5 to 8 ft below land surface.

Robinson (1970) studied ground-water evapotranspiration by woody phreatophytes in the Humboldt River valley near Winnemucca, Nev., from 1963 to 1967. Evapotranspiration tanks were planted with greasewood, rabbitbrush, willow, and wild rose; one tank contained bare soil. Robinson (1970, p. 31-32) concluded, on the basis of tank experiment data, that the consumption of ground water by greasewood ranged from an average of about 0.6 ft/yr to about 0.8 ft/yr from 1963 to 1967. Rabbitbrush transpired an average of about 1.1 ft/yr from 1964 to 1967. Depths to ground water in the greasewood tanks ranged from 5 to 8 ft and in the rabbitbrush tanks, from 5 to 6 ft. These rates are considerably greater than the rates reported by White, but are difficult to compare because of a lack of comparable canopy density and volume data for the two studies.

More recent studies of evapotranspiration by rangeland vegetation have used micrometeorological methods to measure total above-canopy fluxes of sensible and latent heat. Malek and others (1990) measured evapotranspiration from the moist playa and playa margin of Pilot Valley, Utah. Duell (1990) completed similar studies for several different desert shrub and grass communities growing in areas of shallow ground water in Owens Valley, Calif. Czarnecki (1997) measured evapotranspiration at Franklin Lake Playa, Nev., and Weeks and others (1987) studied evapotranspiration by salt cedar, alkali sacaton, kochia, and grass in the Pecos River Valley between Acme and Artesia, N. Mex. Evapotranspiration rates determined by these studies are somewhat less useful in determining ground-water evapotranspiration because they also include evapotranspiration of precipitation and soil moisture as well as evapotranspiration of ground water. Of the studies cited above, comparative plant density and volume information is given only by Duell (1990).

Evapotranspiration from rangeland shrubs was studied at several locations in the central and western Great Basin from 1988 to 1994 (Nichols, 1994). The general applicability of an energy-combination model (Shuttleworth and Gurney, 1990) for partitioning energy budgets, and hence evapotranspiration, between the soil and canopy of sparse-canopy rangeland vegetation was demonstrated using data from central Nevada (Nichols, 1992a). Using the results of this model, calibrated for study sites underlain by shallow ground water in west-central Nevada, estimates were made of ground-water evapotranspiration rates for greasewood, rabbitbrush, and sagebrush ranging from 0.5 to 0.7 ft/yr (Nichols, 1992b, 1993).

The analysis was expanded to include data and estimated transpiration from phreatophytes shrubs from additional sites and a functional relation was developed between ground-water transpiration from phreatophyte shrubs and the depth to ground water, shrub density, and shrub leaf area index (Nichols, 1994). Following field studies of evapotranspiration from saltgrass areas in southern Nevada (Nichols and others, 1997), the analysis was further extended to develop a functional relation between ground-water evapotranspiration and plant cover (Chapter A of this report). Ground-water evapotranspiration estimated by this relation includes evapotranspiration from phreatophyte grasses and shrubs and from associated bare soil. The analysis of the present study uses the functional relation between ground-water evapotranspiration and plant cover (Chapter A) combined with satellite-data-derived vegetation indices (Chapter B) to estimate regional ground-water evapotranspiration.

Recharge Studies

Several methods for estimating regional ground-water recharge from precipitation have been developed over the years and have been summarized by Lerner and others (1990). Ground-water- resources studies in Nevada, mostly eastern Nevada, during the late 1940's, led investigators to conclude that ground-water recharge by precipitation on the valleys of the region could be estimated from knowledge of ground-water discharge by evapotranspiration from phreatophytes on the valley floor. This is most succinctly stated by Maxey and Eakin (1949, p. 40), who developed the method of estimating recharge in Nevada:

> (R)echarge studies ... consisted of estimating the ground-water discharge by natural losses from 13 valleys in east-central Nevada. The recharge for each valley was also estimated using the rainfall-zone map as a basis. The recharge estimates were then balanced by trial-and-error with the discharge estimates.

This excerpt makes clear that ground-water discharge was the controlling factor in developing estimates of recharge from precipitation. Estimates of ground-water discharge by evapotranspiration were assumed reasonable and reliable and the precipitation as derived from the precipitation map of Nevada (Hardman, 1936) also was assumed reasonable and reliable. Two other assumptions, though not explicitly stated, also were made: (1) that many, if not most of these valleys were hydrologically and topographically closed and that all ground water discharged by evapotranspiration from phreatophytes and springs represented the approximate long-term annual discharge and (2) that recharge to the alluvial aquifer beneath the valley was equal to the long-term annual discharge.

Eleven of the 13 valleys in which the recharge method was developed included Ruby, Clover , Independence, Goshute-Antelope (which included a somewhat different area for Antelope Valley than defined by Cardinalli and others (1968) and used herein), Hot Creek, Railroad, Reveille, Kawich, Penoyer, and White River Valleys (pl. 1). The other two valleys included in the original study are uncertain. Watson and others (1976) list 19 valleys reportedly included in the development of the recharge estimation method. The earliest published reports (Maxey and Eakin, 1949; Eakin, 1950; Eakin and others, 1951b) state that the valleys included in the study were in east-central Nevada; therefore, Diamond Valley and Spring Valley (Watson and others, 1976, p. 340) may be the remaining 2 valleys (pl. 1) needed to complete the list of 13, although no reports dating to the 1940's or early 1950's are known for these valleys.

The areas covered by phreatophyte shrubs and grasses were mapped or estimated in the 13 selected valleys of east-central Nevada (Maxey and Eakin, 1949; Eakin and others, 1951b). Using ground-water evapotranspiration values published by Lee (1912) and White (1932), estimates of total ground-water evapotranspiration were made for each of the valleys. The volume of precipitation falling on each valley was estimated for each of the precipitation zones shown on the existing precipitation map of Nevada (Hardman, 1936). The precipitation zones on this map showed areas of precipitation of less than 5 inches, 5 to less than 8 inches, 8 to less than 12 inches, 12 to less than 15 inches, 15 to less than 20 inches, and 20 inches and greater. Precipitation of less than 8 inches was assumed to produce no ground-water recharge. Using the estimated ground-water evapotranspiration for each valley, or group of valleys, as the assumed recharge from precipitation, coefficients by which to multiply the volume of precipitation in each of the remaining four precipitation zones were determined by trial-and-error (Maxey and Eakin, 1949; Eakin and others, 1951b, p. 26, p. 80, p. 151); the coefficients they derived were 0.03, 0.07, 0.15, and 0.25 for the 8-12, 12-15, 15-20, and 20-inch-and-greater zones, respectively. The coefficients were, in effect, the percentage of precipitation in each zone that reached the ground-water reservoir (Maxey and Eakin, 1949, p. 40).

Reconnaissance Ground-Water Studies

The methodologies developed during these early ground-water resources investigations provided the framework for a long-term program to define the ground-water resources of Nevada. These reconnaissance studies began in 1960 (Eakin, 1960) and concluded in 1974 (Van Denburgh and Rush, 1974). The results were a series of reconnaissance reports that describe the water resources of 219 valleys in 60 reports with a strong emphasis in most reports on ground-water budgets. Reconnaissance studies of the water resources of valleys in the present study area (table C3) include those for Newark Valley (Eakin, 1960); Long Valley (Eakin, 1961); Spring Valley (Rush and Kazmi, 1965); Little Fish Lake, Hot Creek, and Little Smoky Valleys (Rush and Everett, 1966); Steptoe Valley (Eakin and others, 1967); Butte Valley (Glancy, 1968); Tippett and Antelope Valleys (Harrill, 1971); and Railroad Valley (Van Denburgh and Rush, 1974). The subsequent allocation of ground-water

Table C3. Previous reconnaissance-level ground-water resources investigations for valleys in eastern Nevada study area

Valley	Reference	Report series and number [1]
Antelope Valley	Eakin and others (1951a)	Bulletin 12
	Harrill (1971)	Reconnaissance 56
Butte Valley	Glancy (1968)	Reconnaissance 49
Clover Valley	Eakin and Maxey (1951b)	Bulletin 12
Goshute Valley	Eakin and others (1951a)	Bulletin 12
Hot Creek Valley	Maxey and Eakin (1951)	Bulletin 12
	Rush and Everett (1966)	Reconnaissance 38
Independence Valley	Eakin and Maxey (1951b)	Bulletin 12
Jakes Valley	Maxey and Eakin (1949)	Bulletin 8
Little Fish Lake Valley	Rush and Everett (1966)	Reconnaissance 38
Little Smoky Valley	Rush and Everett (1966)	Reconnaissance 38
Long Valley	Eakin (1961)	Reconnaissance 3
Newark Valley	Eakin (1960)	Reconnaissance 1
Railroad Valley	Maxey and Eakin (1951)	Bulletin 12
	Van Denburgh and Rush (1974)	Reconnaissance 60
Ruby Valley	Eakin and Maxey (1951a)	Bulletin 12
Spring Valley	Rush and Kazmi (1965)	Reconnaissance 33
Steptoe Valley	Eakin and others (1967)	Reconnaissance 42
Tippett Valley	Harrill (1971)	Reconnaissance 56

[1] Refers to Water Resources Bulletins and Ground-Water Resources Reconnaissance series of reports published by Nevada Department of Conservation and Natural Resources in cooperation with U.S. Geological Survey. See References Cited section for complete citation.

resources by the State of Nevada has been based on the ground-water budgets published in these and other reconnaissance-series reports.

The reconnaissance studies were part of a cooperative program between the State of Nevada and the U.S. Geological Survey to determine the quantity, quality, and distribution of the water resources of the state (Shamberger, 1962). The program, as proposed, included several local and regional studies, research investigations, and data collection activities. As part of this program

> ...the preliminary reconnaissance studies (were) first-stage hydrologic studies... which summarize(d) hydrologic observations. .. These studies suggest the potentials of water development, but only tentatively because of lack of data ... Nevertheless, they provide the basic framework for second, third, and fourth-stage hydrologic studies ...
> (Shamberger, 1962, p. 14).

As such, the ground-water resources reconnaissance studies were never intended to provide definitive ground-water budgets for the areas studied. Rather, the studies were intended to provide a preliminary estimate and guide for more comprehensive future studies given the availability of new data, methods, and technologies.

The present study builds on the hydrologic knowledge provided by the reconnaissance studies and subsequent studies, uses new information on ground-water evapotranspiration determined from field studies from 1988 to 1994, uses new estimates of annual precipitation for Nevada, and utilizes current technologies in the form of satellite-derived remotely sensed data. The present study follows from the earlier reconnaissance studies and provides a new set of estimates of regional ground-water budgets that update the older reconnaissance estimates.

ESTIMATED GROUND-WATER DISCHARGE AND RECHARGE

Ground-water evapotranspiration was estimated as a function of plant cover (Chapter A of this report). Plant cover, in turn, was estimated on a regional scale from a Landsat-data-derived vegetation index (Chapter B). Regional ground-water evapotranspiration then was estimated using the estimated regional plant cover. These estimates of ground-water discharge by evapotranspiration and estimates of interbasin ground-water flow for the valleys in the study area were used to derive estimates of regional ground-water recharge from precipitation. As such, the estimated recharge is not an independent estimate, but is based on the estimated discharge from the valleys.

Coefficients for estimating recharge from precipitation zones of 8 to less than 12 inches, 12 to less than 16 inches, 16 to less than 20 inches, 20 to less than 34 inches, and 34 inches and greater were calculated using a multiple linear regression model based on estimated ground-water evapotranspiration and interbasin flow. The coefficients represent the percentage of precipitation in each zone that recharges the ground-water system. Ground-water budgets for individual valleys were balanced, where needed, by interbasin ground-water flow from or to adjacent valleys. Many occurrences of interbasin flow have been suggested or proposed by earlier studies. Interbasin ground-water flow was not assumed in the absence of supporting geologic, geophysical, or hydrologic evidence.

Discharge by Evapotranspiration

The discharge of ground water by evapotranspiration was estimated for the valleys of the study area using the method described in Chapter B. This method estimates ground-water evapotranspiration as a function of phreatophyte plant cover. Plant cover was estimated from a vegetation index derived from Landsat data using 1985 and 1989 Landsat scenes of the study area. Details of the estimated ground-water evapotranspiration from each valley for 1985 and 1989 are given in table C17 (at the end of this chapter) and by Smith and others (2000). Weighted-average phreatophyte plant cover ranged from an estimated 3.6 percent in Tippett Valley in 1985 to 68.8 percent in a spring-discharge area of Goshute Valley in 1989. Annual ground-water evapotranspiration rates ranged from a minimum of less than 0.1 ft to a maximum of about 2.6 ft. Ground-water evaporation from bare soil and playa areas was estimated to be about 0.15 ft on the basis of field studies in Railroad Valley (M.J. Johnson, U.S. Geological Survey, written commun., 1994), but probably ranges from near zero in areas of deep ground water to as much as 1.25 ft (Stannard, 1997, p. 35) in areas of very shallow ground water (1 ft or less).

The present study estimated summer (May through September), winter (October through April), and annual ground-water evapotranspiration (Chapter A). Winter ground-water evapotranspiration may account for as much as 26 percent of the total annual ground-water evapotranspiration. The early reconnaissance studies estimated what was referred to, in the early reports, as a growing season ground-water evapotranspiration, with no definition of the growing season. Subsequent studies referred to this same evapotranspiration as annual ground-water evapotranspiration. Robinson (1970) went to some length to define a growing season for the phreatophytes of his study and concluded that it extended from sometime in April to sometime in October, implying no ground water evapotranspired from October to April.

However, studies by Duell (1990) and Nichols and others (1997) demonstrated that ground-water evapotranspiration continues through the winter months in areas of shallow ground water. Most of the winter evapotranspiration is evaporation of ground water from bare soils. Some transpiration occurs during the winter months, as defined in this study, because phreatophyte shrubs remain green well into October, and on occasion into November. Shrub growth begins in April. During periods in the winter when soils near the surface are frozen, soil moisture and soil-water vapor derived from ground water continue to move toward the surface and accumulate in response to soil-water tension gradients. When the frozen surface soils thaw, the ground-water-derived soil moisture is evapotranspired.

Evapotranspiration Rates in 1985 and 1989

Ground-water evapotranspiration from the study area was estimated for 1985 and 1989. Precipitation over the study area was below normal (average for the 30-year period 1961-90) during 1985 and 1989 (table C4). Significantly however, 1985 followed 3 years of well above-normal precipitation over most of the study area and 1989 followed 3 years of near- to below-normal precipitation over most of the study area (table C4). Plant cover was estimated for each valley of the study area from a satellite-data-derived vegetation index (Chapter B) and phreatophyte areas in each valley were mapped in the field using U.S. Geological Survey 1:24,000-scale topographic quadrangle maps. The distribution of phreatophyte plant cover for 1985 is shown on plate 2 and for 1989 on plate 3. These plates also show the distribution of estimated ground-water evapotranspiration because it is directly related to plant cover (Chapter A). The generally wetter conditions preceding 1985 are indicated by water on the playas of Newark, Steptoe, and Spring Valleys, in Franklin Lake and Ruby Lake in Ruby Valley, and in Snow Water Lake in Clover Valley (pl. 2). These lakes, except for Ruby Lake, had dried up by 1989 (pl. 3).

Interestingly though, plant cover was somewhat greater in many of the valleys in 1989 than in 1985. This is most readily seen by comparing the plant cover in 1989 in Goshute, Independence, Butte, and Steptoe Valleys (pl. 3, table C17) with the plant cover in these same valleys in 1985 (pl. 2, table C17). Plant cover was about the same in both years in Railroad, Little Fish Lake, Hot Creek, Little Smoky, and Newark Valleys. Plant cover in Spring Valley was less in 1989 than in 1985. These changes, however, were caused by rather small changes in the amount of plant cover in areas where the cover is 20 percent or less. These areas are covered by phreatophyte shrubs, the numbers or density of which did not change between 1985 and 1989.

Table C4. Mean annual and annual precipitation for selected stations and time periods, eastern Nevada. Station locations shown on plate 2

U.S. Weather Station name	Mean annual precipitation (inches)			Annual precipitation (inches)	
	1961–90	1982-84	1986–88	1985	1989
Elko	10.13	15.07	9.85	9.89	6.60
Ely	9.01	11.03	8.97	6.98	6.18
Eureka	10.56	16.08	8.20	7.89	8.66
Lund	13.10	19.15	11.12	6.82	7.21
Ruby Lake	9.95	14.14	7.14	7.30	7.88
Snowball Ranch	13.22	19.44	10.85	10.84	10.28
Wells	10.46	16.38	10.29	11.22	7.46

The changes were caused solely by increases or decreases in shrub leaf area index, *LAIp*. For example, an area in 1985 with a shrub density, *d*, of 0.20 (20 percent) and shrub leaf area index, *LAIp*, of 1.5, would have a plant cover, *Cp*, of 0.075 (7.5 percent), which places such an area in the less-than-10 percent plant cover zone. This same area in 1989, with the same shrub density, *d*, of 0.20 (20 percent) but with a shrub leaf area index, *LAIp*, of 2.5 would have a *Cp* of 0.125 (12.5 percent), which would move the area from the less-than-10 percent plant cover zone into the 10-20 percent plant cover zone. This change is readily seen in the data given in table C17 at the end of this chapter.

In Goshute Valley in 1985, for example, 4,197 acres were classified as bare soil, 129,981 acres with less than 10 percent plant cover, and 1,178 acres with 10-20 percent cover. In 1989, 4,156 acres that previously had such sparse vegetation that they were considered bare soil now had enough vegetation to be moved into the less than 10 percent zone while the leaf area of shrubs on 11,165 acres of the less than 10 percent zone had increased enough to move these areas into the 10-20 percent zone (fig. C3). The average plant cover increased slightly in the less than 10 percent zone from 4.7 percent in 1985 to 7.8 percent in 1989, but decreased slightly in the 10-20 percent zone from 13.9 percent in 1985 to 12.5 percent in 1989.

The reasons for these small, but significant, changes in shrub leaf area index are uncertain. Several possible explanations, however, are offered. The higher-than-normal precipitation during the 3 years before 1985 may have left the soils in some of the lower areas of the valley floors wet enough that it inhibited spring growth of greasewood, which is the dominant shrub in the areas that exhibited most of the change in plant cover. Groeneveld (1989) reported significant dieback by greasewood under flooded conditions. Robertson (1983, p. 315) also reported that greasewood dies when the water table is at land surface and White (1932, p. 33) suggested that greasewood preferred about 3 ft of aerated soil.

Another explanation for the difference in leaf area between the 2 years may be related to nutrients. Groeneveld (1989) has suggested a correlation between the depth distribution of greasewood roots and the availability of nitrogen, presumably from plant detritus accumulated at land surface. The availability of nitrogen from this source also may be a factor in leaf development. If so, then this source of nitrogen may have been depleted by the higher-than-normal precipitation during the 3 years before 1985, but may not have been depleted by the below-normal precipitation during the 3 years before 1989. Factors other than these, or in addition to these, may be responsible for the differences in leaf area observed in 1985 and 1989; determination of these factors by other investigations may be important in understanding the dynamics of rangeland vegetation, but are beyond the scope of this work. Estimated ground-water evapotranspiration from the valleys of the study area in 1985 and 1989, and the estimated mean annual ground-water evapotranspiration from each valley is summarized in table C5.

Ground-water evapotranspiration from specific land-cover zones, the estimated weighted-mean plant cover for the zone, the area of the zone, and annual estimated ground-water evapotranspiration rate are summarized in table C6; details for each valley for each year, including the estimated summer (May through September), winter (October through April), and annual ground-water evapotranspiration are given in table C17 at the end of this chapter. Slightly more than 13 percent of the study area was covered by phreatophytes, bare soil, or, in 1985, water-covered playas and lakes (table C6). Areas with plant cover equal to or greater than 20 percent comprised only about 9 percent of the total phreatophyte area in 1985 and about 8 percent in 1989; the remaining

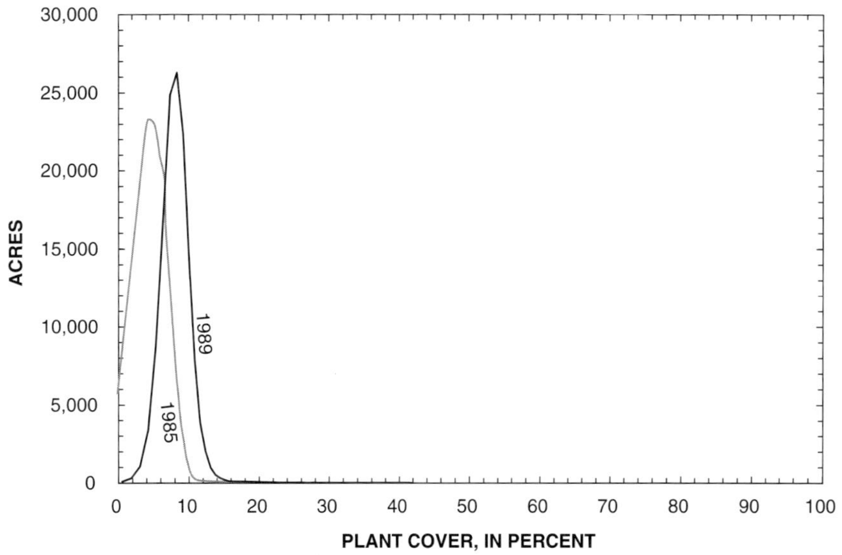

Figure C3. Distribution of plant cover in Goshute Valley, 1985 and 1989.

Table C5. Estimated ground-water evapotranspiration from phreatophytes, bare soil, and open water, eastern Nevada study area, for 1985, 1989, and mean annual conditions [1]

Valley	Estimated ground-water evapotranspiration (acre-feet)		
	1985	1989	Mean annual
Antelope Valley	1,900	6,200	4,000
Butte Valley	37,000	60,000	44,500
Clover Valley	82,000	87,000	84,500
Goshute Valley	28,500	83,500	42,500
Hot Creek Valley	5,000	4,900	5,000
Independence Valley	42,500	63,500	47,000
Jakes Valley	600	500	600
Little Fish Lake Valley	11,000	10,000	9,700
Little Smoky Valley	5,400	6,700	6,000
Long Valley	9,200	12,500	11,000
Newark Valley	59,000	62,000	60,500
Railroad Valley (northern part)	86,500	83,500	85,000
Ruby Valley	[2]157,000	[2]177,000	[2] 167,000
Spring Valley	102,000	77,500	90,000
Steptoe Valley	118,000	137,000	128,000
Tippett Valley	900	4,900	2,900
Total	746,000	877,000	788,000

[1] Numbers are rounded to the nearest 100 acre-ft for values less than 10,000 acre-ft, to the nearest 500 acre-ft for values equal to or greater than 10,000 acre-ft and less than 100,000 acre-ft, and to the nearest 1,000 acre-ft for values equal to or greater than 100,000 acre-ft.

[2] Includes 22,000 acre-ft estimated from Ruby Lake Marsh. See discussion in text.

91 to 92 percent of the phreatophyte area was comprised of areas with plant cover less than 20 percent, bare soil, playa, or water-covered playa. The area of permanent water was insignificant.

Areas with 20 percent or more plant cover accounted for about 36 percent of estimated ground-water evapotranspiration in 1985 and about 27 percent in 1989 (table C6). The estimated mean annual rate of ground-water evapotranspiration was two to two and a half times greater than the rate for these areas used by the earlier reconnaissance series of studies in 1985 and 1989. Areas with less than 20 percent plant cover accounted for about 64 percent of estimated ground-water evapotranspiration in 1985 and about 73 percent in 1989. The estimated mean annual rate of evapotranspiration was about 0.3 ft in 1985 for those areas with less than 10 percent plant cover, similar to the rate used by many of the earlier reconnaissance studies, and about 1.3 ft for areas of 10-20 percent plant cover in 1985 (table C6). Estimated rates were similar for 1989.

Estimated Mean Annual Ground-Water Evapotranspiration

A mean annual value of ground-water evapotranspiration implies a hydrologic steady-state condition. This, in turn, implies no change in ground-water storage and no change in interbasin ground-water flow. When estimates of annual ground-water evapotranspiration differ from one year to another (table C5), changes in ground-water storage, as well as interbasin ground-water flow, should be included in any method used to estimate long-term mean annual ground-water discharge. A rise in ground-water levels over an annual cycle indicates an excess of recharge to the valley-fill aquifer over discharge from the aquifer, but implies nothing about when the recharge occurred in the adjacent mountain block. In this case, recharge may have been greater than the long-term mean annual recharge or discharge may have been less than the long-term mean annual discharge, or more likely a combination of both. Similarly, a decline in ground-water levels over an annual cycle indicates an excess of discharge

Table C6. Summary of area, percent plant cover, mean annual ground-water evapotranspiration (ET) rate, and annual ground-water evapotranspiration for indicated land cover for 1985 and 1989 for eastern Nevada study area [1]

[Symbols: <, less than; ≥, greater than or equal to]

Land cover	1985				1989			
	Area (acres)	Plant cover (percent)	Annual ET rate (feet)	Annual ET (acre-feet)	Area (acres)	Plant cover (percent)	Annual ET rate (feet)	Annual ET (acre-feet)
Water	82,431			650	1,487			650
Bare soil/playa	162,736		0.150	24,410	122,341		0.150	18,351
< 10 percent plant cover	728,988	6.0	.290	211,361	804,624	7.1	.410	322,891
10 – < 20 percent plant cover	171,415	13.7	1.346	230,777	227,276	13.1	1.276	290,003
20 – < 35 percent plant cover	67,745	25.3	2.144	145,252	58,178	25.6	2.154	125,291
35 – < 50 percent plant cover	23,980	41.2	2.506	60,103	20,902	40.6	2.504	52,346
≥ 50 percent plant cover	20,697	63.1	2.584	53,485	23,184	62.9	2.582	59,847
Total [2]	1,258,000			726,000	1,258,000			869,000
Weighted average		[3].105	.577			[3].110	.690	

[1] Calculated values are not rounded, to minimize rounding errors in subsequent calculations, except as indicated.

[2] Totals are rounded to the nearest 1,000 acres and acre-ft.

[3] Does not include areas of bare soil, playa, or areas covered by water.

over recharge, which may represent discharge greater than the long-term mean annual discharge or recharge less than the long-term mean annual recharge, or again a combination of both. Interannual changes in ground-water levels commonly are a foot or two and as such do not significantly increase or decrease water-level gradients, and consequently ground-water flow rates, in areas of interbasin ground-water flow. Annual ground-water evapotranspiration in combination with annual ground-water recharge is largely responsible for changes in ground-water storage and, therefore, changes in ground-water levels.

Appropriate ground-water level data are sparse in the valleys of the study area, and detailed estimates of changes in ground-water storage were not possible. Only Ruby, Clover, and Steptoe Valleys have water-level data that provide some insight into changes in ground-water storage in 1985 and 1989. Even in these valleys, however, data were insufficient to characterize the change in storage over the entire valley. While including the change in ground-water storage together with ground-water evapotranspiration is an objective method for estimating long-term mean annual discharge, this method cannot be applied in the present study because of the lack of appropriate water-level data. Consideration of this method should be given in future studies of this type.

Ground-water evapotranspiration estimated for 1985 was similar to that estimated for 1989 in Clover, Hot Creek, Jakes, Little Smoky, Long, Newark, and Railroad Valleys; the average of the 2 years was assumed to approximate mean annual ground-water evapotranspiration in these valleys. Ground-water evapotranspiration estimated for 1989 from Antelope and Tippett Valleys exceeded the estimates for 1985, but the totals for each year are relatively small; the average of the 2 years was assumed to approximate the mean annual value for these two valleys as well. Differences between ground-water evapotranspiration estimated for 1985 and 1989 for Ruby, Spring, and Steptoe Valleys are substantial, but in the absence of sufficient water-level data with which to estimate changes in ground-water storage to reconcile the difference, the average of the 2 years was assumed to approximate mean annual ground-water evapotranspiration. Ground-water evapotranspiration estimated for 1989 from Butte, Goshute, and Independence Valleys significantly exceeded that estimated for these valleys for 1985. Estimates of mean annual ground-water evapotranspiration for these valleys, and for Little Fish Lake Valley (table C5), were developed during determination of mean annual recharge estimates discussed below.

The depth to ground water is greater than 100 ft beneath Jakes, Tippet, and much of Antelope Valleys. Previous studies did not include evapotranspiration from the phreatophytes growing in these areas in the ground-water budget. Field studies as part of this investigation and elsewhere in Nevada (Nichols, 1994) have shown that phreatophytes, even in areas with depths to ground water of no more than 20 ft, may subsist on perched ground water at shallower depths. Phreatophytes in Tippett Valley and those areas of deep ground water in Antelope Valley are assumed to be supported by perched ground water, and while this ground water cannot be developed, it is water that is a part of the ground-water budget. No ground-water evaporation is estimated from bare soil or playa areas of these two valleys. Ground-water evapotranspiration in Jakes Valley is from phreatophytes in the valley of Illipah Creek, which is tributary to Jakes Valley. No ground-water evaporation is estimated from the small area of bare soil or playa in the main valley.

The marshes of the Ruby Lake National Wildlife Refuge in southern Ruby Valley present a particular problem for the approach used in this study. Plant cover values for the marsh area are not correct because the vegetation index derived from Landsat data is incorrect, although water areas are properly identified. Much of the green vegetation in the marsh area during mid to late June is obscured or partly obscured from the satellite's view by senescent vegetation from the previous year(s). The preponderance of senescent vegetation returns reflectance values to the satellite sensor that are in the range for bare soil. This was especially true in 1985 when the plant cover data indicated that about 70 percent of the marsh was bare soil, when, in fact, bare soil probably was not visible in the marsh area because of the large area of open water in 1985 from previous years of above-normal precipitation. Plant cover data for 1989 indicate about 27 percent of the marsh area was bare soil; probably about 17 percent of the marsh area, limited to the northeastern and eastern border of the marsh was bare soil in 1989.

The marsh area in southern Ruby Valley covers about 14,600 acres. In 1985, open water within the marsh area covered 3,535 acres. By 1989, after 3 years of below-normal precipitation and 1 year of near-normal precipitation, the open water area of the marsh was 1,030 acres. To estimate evapotranspiration from the marsh area, the water area of 1989 is assumed to represent the approximate mean annual area of open water. The rest of the marsh area, under mean annual conditions, includes about 11,100 acres of vegetation of all types and about 2,500 acres of bare soil, the area along the northeastern and eastern borders of the marsh covered by water in 1985. Estimated open-water annual evaporation in this area of Nevada is about 4 ft (Scott, 1971) suggesting evaporation from open water of about 4,120 acre-ft/yr. Evapotranspiration from marsh vegetation may range from 1.5 to 3.5 ft/yr (G.A. DeMeo, U.S. Geological Survey, written commun., 1998); a mean annual estimate of about 2 ft is assumed for this analysis and suggests evapotranspiration of 22,200 acre-ft/yr from all vegetation within the marsh area. About 500 acre-ft of ground water may evaporate from the bare soil area. Total estimated evapotranspiration from water, vegetation, and bare soil in the marsh area is estimated to be about 26,800 acre-ft/yr, which includes surface water inflow to the marsh area as well as ground water. Estimates are that perhaps 5,000 acre-ft/yr of this is supported by surface-water inflow and that about 22,000 acre-ft/yr is supported by ground water. This was added to the ground-water evapotranspiration of about 135,000 acre-ft/yr in 1985 and about 155,000 acre-ft/yr in 1989 estimated from plant cover data for the rest of Ruby Valley (tables C5 and C17).

Mean annual ground-water evapotranspiration estimated by this study is compared with that estimated by the reconnaissance series of studies in table C7. The present study estimates slightly more than twice as much ground-water evapotranspiration from the valleys of the study area than these earlier studies.

Recharge From Precipitation

Estimates of ground-water discharge by evapotranspiration from the valleys of the study area provide a basis for estimating ground-water recharge from precipitation; any additional discharge from or recharge to a valley through interbasin ground-water flow must be estimated or known independently. The estimates of mean annual ground-water evapotranspiration developed by this study and estimates of regional interbasin ground-water flow suggested by previous studies summarized and compiled by Harrill and others (1988) were used to determine new estimates of ground-water recharge to the valleys of the study area using a map of annual precipitation in Nevada for 1961-90.

Table C7. Comparison of estimated mean annual ground-water evapotranspiration for eastern Nevada study area, from reconnaissance studies and from this study

Valley	Mean annual ground-water evapotranspiration (acre-feet)	
	Reconnaissance studies	This study [1]
Antelope Valley	100	4,000
Butte Valley	19,900	44,500
Clover Valley	19,000	84,500
Goshute Valley [2]	10,075	42,500
Hot Creek Valley	4,600	5,000
Independence Valley	9,500	47,000
Jakes Valley	--	600
Little Fish Lake Valley	10,000	9,700
Little Smoky Valley	1,900	6,000
Long Valley	2,200	11,000
Newark Valley	16,000	60,500
Railroad Valley (northern part) [3]	80,000	85,000
Ruby Valley	67,600	167,000
Spring Valley	70,000	90,000
Steptoe Valley	70,000	128,000
Tippett Valley	0	2,900
Total	[1]381,000	788,000

[1] Numbers are rounded to the nearest 100 acre-ft for values less than 10,000 acre-ft, to the nearest 500 acre-ft for values equal to or greater than 10,000 acre-ft and less than 100,000 acre-ft, and to the nearest 1,000 acre-ft for values equal to or greater than 100,000 acre-ft.

[2] Reconnaissance estimate includes estimate for Antelope Valley.

[3] Van Denburgh and Rush (1974).

Regional Precipitation

Regional precipitation is difficult to characterize for simple terrain conditions, but those difficulties are compounded by orographic effects of the complex terrain of the intermountain west including Nevada and the Great Basin. Multiple storm tracks and rainshadow effects preclude a simple relation between precipitation and local altitude. The western and southern parts of the study area, including Newark, Little Fish Lake, Little Smoky, Hot Creek, and Railroad Valleys, are influenced by a westerly winter storm track, but lie in the rainshadow of the Sierra Nevada in western Nevada and eastern California, and the Toiyabe Range in central Nevada (fig. C1). The northern and eastern part of the study area is affected, to a greater extent, by a northwesterly storm track. Storms moving southwest over the low terrain of southern Idaho and north-central Nevada do not reach high-altitude topography until the Ruby Mountains and East Humboldt Ranges that form the northwestern border of the study area. These ranges, consequently, receive significantly greater precipitation than most other mountain ranges in Nevada. The rainshadow effects of the Ruby Mountains and the Toiyabe Range can be seen in the precipitation record (table C8) for ranges from the Ruby mountains to the southeast (pl. 1). The precipitation-altitude relation demonstrated by these data (fig. C4) is significantly different over most of the study area from the relation assumed by the reconnaissance studies.

A model for estimating and distributing annual average precipitation at a regional scale that is well suited to areas with mountainous terrain has recently been developed (Daly and others, 1994). The model, known by the acronym PRISM (Precipitation-elevation Regression on Independent Slopes Model), was used to generate a map of estimated annual precipitation for Nevada for 1961-90 (G.H. Taylor, Oregon State University, written commun., May 21, 1997). The annual precipitation estimates on this map were used, in this study, to estimate ground-water recharge to the eastern Nevada study area.

Annual precipitation in the valleys of the study area estimated by the PRISM model is compared in table C9 to that estimated from the Hardman (1936) map and used by the earlier reconnaissance studies. Precipitation estimated from the PRISM data exceeds that estimated from the Hardman map for the reconnaissance studies. Increases ranged from 101 percent greater than the Hardman map estimate in Steptoe Valley to 141 percent greater in Antelope Valley. For estimating recharge, a more significant comparison is of precipitation in areas that receive annual precipitation greater than 20 inches. The area, volume, and estimated average precipitation greater than 20 inches from the reconnaissance studies and the area, volume, and weighted average of precipitation calculated from the PRISM data are given in table C10. Significantly more precipitation is estimated by PRISM in the greater-than-20-inch zone for Butte, Clover, Independence, Newark, Railroad, Ruby, Spring, and Steptoe Valleys than was estimated by the reconnaissance studies. The increase in volume ranged from about 200 to 300 percent for most of the valleys to more than a 1,000 percent increase in Butte and Newark Valleys.

Table C8. Precipitation at selected SNOTEL stations in eastern Nevada, 1961-90 (Greenlee, 1992)

Station name [1]	Mountain range	Altitude (feet above sea level)	Mean annual precipitation (inches)
Lamoille #3	Ruby Mountains	7,700	33.9
Hole-In-Mountain	East Humboldt Range	7,900	33.6
Green Mountain	Ruby Mountains	8,000	32.9
Dorsey Basin	Ruby Mountains	8,100	31.9
Corral Canyon	Ruby Mountains	8,500	29.8
Diamond Peak	Diamond Mountains	8,000	24.6
Berry Creek	Schell Creek Range	9,100	26.0
Ward Mountain	Egan Range	9,200	24.7

[1] Locations are shown on plate 1.

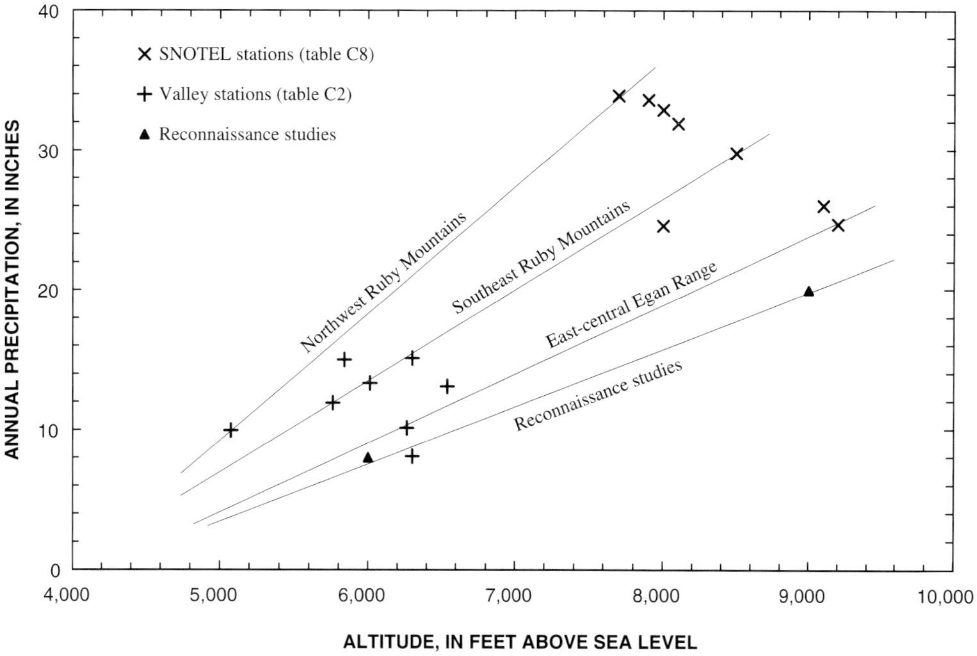

Figure C4. Precipitation-altitude relation, eastern Nevada.

The May 1997 version of the PRISM map of mean annual precipitation of Nevada for 1961-90 was used to estimate precipitation in the valleys of the study area. One version of this map portrays precipitation amounts in 1-inch contour intervals from 4 to 22 inches, 2-inch intervals from 22 to 26 inches, 4-inch intervals from 26 to 42 inches, and 2-inch intervals for greater than 42 inches. A second version uses 2-inch contour intervals from 4 to 20 inches and a 4-inch contour interval for greater than 20 inches. These data were used to create a map using 1-inch contour intervals over the entire range of precipitation values.

Determination of Recharge Coefficients

The areas and volumes corresponding to 1-inch increments of mean annual precipitation were determined for each valley of the study area (table C18 at the end of this chapter), which then were combined into larger intervals for use in recharge calculations. Following Maxey and Eakin (1949) and Eakin and others (1951b), no recharge was assumed from precipitation of less than 8 inches. The remaining precipitation zones were combined into the following intervals: 8 to less than 12 inches, 12 to less than 16 inches, 16 to less than 20 inches, 20 to less than 34 inches, and equal to or

Table C9. Area and annual volume of precipitation estimated by reconnaissance studies and by present study for valleys in eastern Nevada study area [1]

[Area is in acres and volume of precipitation is in acre-feet per year. Symbol: --, not reported.]

Valley	Reconnaissance studies		This study (PRISM)	
	Area	Precipitation	Area	Precipitation
Antelope Valley	252,600	175,200	255,680	246,551
Butte Valley	635,400	563,300	652,362	700,905
Clover Valley	288,100	287,870	292,115	363,328
Goshute Valley	--	--	612,168	592,875
Hot Creek Valley	657,990	397,340	658,500	424,067
Independence Valley	336,000	296,280	360,670	394,414
Jakes Valley	--	--	270,498	289,477
Little Fish Lake Valley	278,260	226,750	276,483	236,430
Little Smoky Valley	743,840	432,930	740,576	523,359
Long Valley	416,000	343,940	419,844	452,367
Newark Valley	512,000	410,490	509,283	515,471
Railroad Valley (northern part)	1,376,800	996,000	1,369,671	1,089,249
Ruby Valley	639,900	682,550	638,935	867,225
Spring Valley	1,084,900	962,790	1,067,010	1,141,444
Steptoe Valley	1,265,000	1,328,310	1,245,618	1,344,191
Tippett Valley	233,000	164,500	221,574	211,905

[1] Calculated values are not rounded, to minimize rounding errors in subsequent calculations.

Table C10. Comparison of area, average annual rate, and annual volume of precipitation greater than 20 inches estimated by reconnaissance studies and equal to or greater than 20 inches estimated by the present study for valleys in eastern Nevada study area

[Area is in acres, mean precipitation is in inches per year, and volume is in acre-feet per year.]

Valley	Reconnaissance studies			Present study		
	Area, >20 inches	Mean precipitation	Volume	Area, ≥20 inches [1]	Weighted mean precipitation	Volume[1]
Antelope Valley	0	--	0	520	20.0	867
Butte Valley	4,300	21.6	7,740	42,607	22.4	79,414
Clover Valley	15,600	25.0	32,448	38,542	26.7	85,764
Goshute Valley	0	--	0	5,584	20.8	9,675
Hot Creek Valley	0	--	0	570	20.0	950
Independence Valley	0	--	0	9,251	20.7	15,970
Jakes Valley	0	--	0	607	20.0	1,011
Little Fish Lake Valley	2,390	21.0	4,183	0	--	0
Little Smoky Valley	0	--	0	3,407	20.0	5,678
Long Valley	0	--	0	5,604	21.5	10,050
Newark Valley	3,000	21.0	5,250	30,283	22.1	56,210
Railroad Valley (northern part)	22,000	21.6	39,600	48,281	22.3	89,567
Ruby Valley	58,900	25.0	122,512	109,401	26.9	244,863
Spring Valley	59,100	21.0	103,425	96,091	24.3	194,247
Steptoe Valley	57,000	21.0	99,750	120,169	23.6	236,437
Tippett Valley	280	21.6	504	958	20.0	1,596

[1] Calculated values are not rounded, to minimize rounding errors in subsequent calculations.

greater than 34 inches. These intervals were decided upon after solving for recharge coefficients, or percentages, using several combinations of intervals and are similar to those used by Maxey and Eakin (1949) and Eakin and others (1951b) with the exception of the zone with precipitation greater than 34 inches.

Two criteria were used in determining recharge coefficients, or percentages:

1. No coefficient was assumed to be less than 0 because a negative percentage has no physical meaning, and
2. The coefficients, or percentage of recharge, should increase from the smallest value for the 8-to-less-than-12-inch zone to the largest value for the equal-to-or-greater-than-34-inch zone.

With the volume of precipitation for each precipitation zone and an estimate of ground-water recharge from precipitation based on the estimated ground-water discharge, the following multiple-linear regression model can be solved (see also Watson and others, 1976):

$$Y = b_0 + \sum_{n=1}^{5} b_i X_i + \varepsilon_i \qquad (1)$$

where Y is the estimated recharge, in acre-feet, based on the estimated discharge, in acre-feet;

X_i is the independent variable, in this case precipitation volume, in acre-feet, in each of the five precipitation zones;

b_i is the coefficient for each independent variable;

b_0 is the intercept, in acre-feet, on the Y axis; and

ε_i is the error, in acre-feet, in the estimated discharge.

Estimated ground-water discharge by evapotranspiration must be corrected for any interbasin flow into or from the basin to reflect the total recharge needed to maintain steady-state equilibrium of the ground-water system in the valley. Correctly defining recharge and discharge is further complicated by the assumption that the hydrologic basin is coincident with the topographic basin; that is, hydrologic divides are defined by topographic divides. The exact locations of ground-water divides in the study area are not known, and in the fractured and faulted carbonate-rock terrain of eastern Nevada these divides may be substantially displaced from topographic divides. The present study followed conventional practice and assumed that the hydrologic basin is coincident with the topographic basin; any contribution to or loss from another topographic basin was assumed to occur as interbasin ground-water flow.

Estimated ground-water discharge by evapotranspiration was modified by adding or subtracting estimated interbasin ground-water flow suggested by previous studies (summarized by Harrill and others, 1988) and by analysis of the distribution and location of ground-water evapotranspiration estimated by this study. Application of the estimated magnitude of interbasin flow summarized by Harrill and others (1988) is problematic, because these magnitudes are based on the reconnaissance estimates of ground-water discharge and recharge. However, the locations and directions of interbasin flow (fig. C5) given by Harrill and others (1988) provide a framework within which to determine new estimates of the magnitude of interbasin flow.

Analysis of the distribution of ground-water evapotranspiration estimated by the present study provides additional information about the magnitude of interbasin flow for several valleys. Previous studies have suggested that ground water discharged by springs near Duckwater and near Lockes in Railroad Valley (pl. 1) represents interbasin flow from adjacent valleys (Rush and Everett, 1966; Van Denburgh and Rush, 1974; Prudic and others, 1995). The present study estimated ground-water evapotranspiration of 12,500 acre-ft/yr in the Duckwater area and as much as 9,000 acre-ft/yr in the Lockes area (Chapter B, table B2) suggesting that as much as 21,500 acre-ft/yr of ground-water evapotranspiration in Railroad Valley may not be derived from recharge by precipitation in the valley.

Interbasin flow to Newark Valley from Long and Little Smoky Valleys has been suggested previously by Harrill and others (1988). Examination of the phreatophyte plant cover distribution (pls. 2 and 3) suggests that the phreatophytes covering about 2,200 acres in southwestern Newark Valley probably are related to interbasin flow derived from spring discharge at Fish Creek Springs (pl. 1) in Little Smoky Valley. The large area of phreatophytes in northeastern Newark Valley (pls. 2 and 3), in the area of Warm Springs Ranch (pl. 1), may well be supported by ground water moving from Long Valley through the thick sequence of Devonian carbonate rocks underlying the mountains to the east; between 2,500 and 6,500 acres of phreatophytes

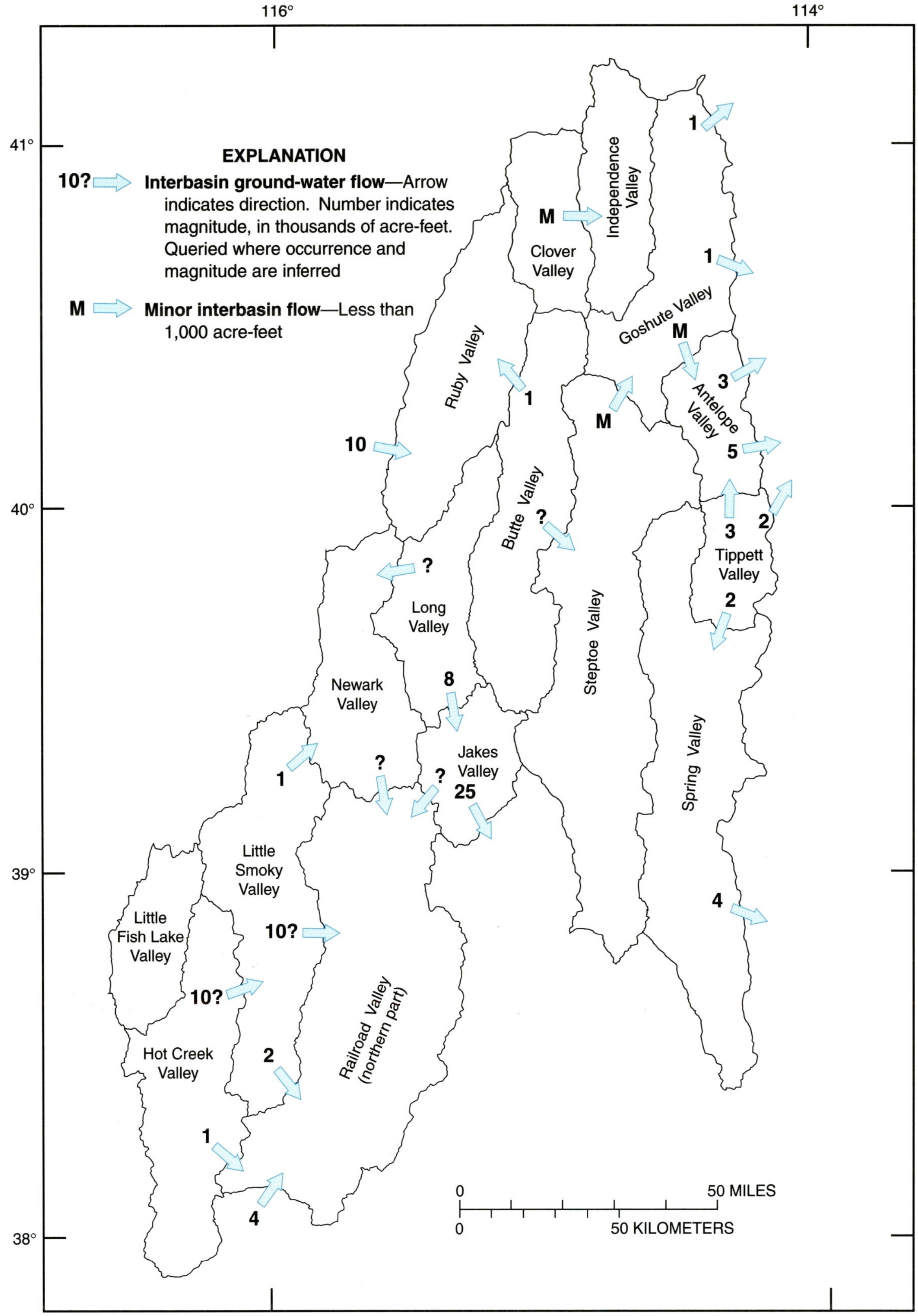

Figure C5. Directions, locations, and estimated magnitude of interbasin ground-water flow in eastern Nevada as suggested by earlier studies; from Harrill and others (1988) except northward inflow to northern part of Railroad Valley, which is from Van Denburgh and Rush (1974). Directions, locations, and estimated magnitude of interbasin ground-water flow in eastern Nevada based on the results of the present study are shown on plate 4.

may be supported by this interbasin flow. Similarly, as much as 8,500 acres of phreatophytes in the southeastern phreatophyte area in central Newark Valley south of Buck Mountain (pls. 2 and 3) may be supported by interbasin flow from Long Valley through the upper Paleozoic carbonate rocks and Tertiary volcanic rocks underlying the mountains to the east of Newark Valley in this area. Estimated ground-water evapotranspiration from these areas, from the present analysis, is about 1,500 acre-ft/yr in southeastern Newark Valley, from about 5,000 to about 9,000 acre-ft/yr from the Warm Springs Ranch area, depending on the size of the area supported by interbasin flow, and as much as 6,000 acre-ft/yr from the southeastern phreatophyte area south of Buck Mountain, again depending on the area supported by interbasin flow. This suggests that as much as 16,500 acre-ft/yr of ground-water evapotranspiration in Newark Valley may be derived from recharge from precipitation outside the topographic limits of the valley.

The multiple linear regression model (eq. 1) was used to derive new coefficients, or percentages, by which to estimate ground-water recharge from precipitation. Ground-water evapotranspiration from each valley estimated by the present study (table C7) and modified by interbasin ground-water flow estimated by this study or by previous studies, as summarized by Harrill and others (1988), were used in the model as the initial estimates of ground-water recharge from precipitation to each basin. Ground-water evapotranspiration for each valley and estimated values of probable and proposed net interbasin ground-water flow are given in table C11.

Probable values of predicted interbasin flow are based on hydrologic measurements, such as spring discharge, or on ground-water evapotranspiration

Table C11. Estimated and predicted ground-water evapotranspiration and interbasin ground-water flow for predicting ground-water recharge to valleys in eastern Nevada study area [1]

[All values in acre-feet per year]

Valley	Ground-water evapotranspiration from this study: Initial estimate	Net interbasin ground-water flow [2]		Estimated total ground-water discharge for regression analysis (A + C)	Predicted ground-water recharge from regression analysis, equation 1	Predicted total ground-water discharge from regression analysis	Ground-water evapotranspiration adjusted from regression analysis (table C5)	Predicted net interbasin flow from regression analysis[2] (E - G)	Difference between predicted recharge and estimated discharge (E - D)
		From previous studies (fig. C5)	For regression analysis, this study						
	A	B	C	D	E	F	G	H	I
Antelope Valley	4,000	+5,000	+12,000	16,000	16,872	16,872	4,000	+12,872	+872
Butte Valley	47,000	+1,000	+23,500	70,500	69,122	69,122	44,500	+24,622	-1,378
Clover Valley	84,500	0	-25,400	59,100	58,872	58,872	84,500	-25,628	-228
Goshute Valley	42,500	+2,000	-1,500	41,000	41,026	41,026	42,500	-1,474	+26
Hot Creek Valley	5,000	+11,000	+800	5,800	5,806	5,806	5,000	+806	+6
Independence Valley	48,000	0	+2,200	50,200	50,142	50,142	47,000	+3,142	-58
Jakes Valley	600	+17,000	+17,000	17,600	38,259	38,259	600	+37,659	+20,659
Little Fish Lake Valley	9,600	0	0	9,600	9,674	9,674	9,700	-26	+74
Little Smoky Valley	6,000	+3,000	+5,500	11,500	12,753	12,753	6,000	+6,753	+1,253
Long Valley	11,000	+8,000	+27,000	38,000	47,826	47,826	11,000	+36,826	+9,826
Newark Valley	60,500	-1,000	-8,500	52,000	49,189	49,189	60,500	-11,311	-2,811
Railroad Valley (northern part)	85,000	-17,000	-18,000	67,000	61,234	61,234	85,000	-23,766	-5,766
Ruby Valley	148,000	-11,000	0	148,000	145,795	145,795	167,000	-21,205	-2,205
Spring Valley	90,000	+2,000	+4,000	94,000	103,777	103,777	90,000	+13,777	+9,777
Steptoe Valley	127,500	0	+2,000	129,500	131,716	131,716	128,000	+3,716	+2,216
Tippett Valley	2,900	+7,000	+9,000	11,900	12,430	12,430	2,900	+9,530	+530
Total	772,100	+27,000	+49,600	821,700	854,493	854,493	788,200	+66,293	+32,793

[1] Calculated values are not rounded, to minimize rounding errors in subsequent calculations.

[2] Positive value is estimated net flow from indicated basin. Negative value is estimated net flow into indicated basin.

estimated by the present study, such as those described for Railroad and Newark Valleys; these are described more fully in the following section. Proposed values of predicted interbasin ground-water flow are not demonstrated by direct hydrologic measurement. They are, however, supported by geologic and hydrologic conditions that are described more fully in the following section. The model then was calibrated by increasing or decreasing estimated interbasin flow, where necessary, for predicted recharge to equal estimated discharge for each basin, while also meeting the two criteria given above.

Data for 14 of 16 basins were used to solve the multiple linear regression model (eq. 1); Jakes Valley was excluded from the analysis because essentially no ground water evapotranspired from the valley. Data for Goshute Valley were excluded during preliminary calibration of the model because of the large difference between the 1985 and 1989 estimated ground-water evapotranspiration. Initial solutions of the model suggested recharge to Goshute Valley of about 41,000 acre-ft/yr. An estimated ground-water evapotranspiration of 42,500 acre-ft/yr was selected to be consistent with a balanced ground-water budget that also accommodated previous estimates of interbasin flow (Harrill and others, 1988) and excess recharge to Steptoe Valley estimated by the present study. The data for Clover, Independence, and Butte Valleys were combined and treated as data for a single valley for initial model calibration. Data for each valley separately were used for final model calibration. Estimated mean annual ground-water evapotranspiration for Independence and Butte Valleys were determined so as to allow sufficient interbasin flow to satisfy ground-water evapotranspiration from Clover Valley. The predicted interbasin flows are consistent with hydrologic conditions as discussed below.

The coefficients, or percentages, by which to multiply precipitation in each precipitation zone are given in table C12. The y-axis intercept, b_0, for equation 1 was 8.7; setting b_0 to zero leads to no change in the coefficients. Statistics for the regression model are not valid because it was calibrated so that estimated recharge equaled total estimated discharge for each valley and the solution therefore has an $r^2 = 1.0$ (table C11). However, simple linear regression of estimated discharge (Column D, table C11) against predicted recharge (Column E, table C11) yields an r^2 of 0.975 and an adjusted r^2 of 0.909.

Table C12. Coefficients for estimating recharge from precipitation in eastern Nevada study area

Precipitation zone (inches)	Coefficient
8 to less than 12	0.008
12 to less than 16	.130
16 to less than 20	.144
20 to less than 34	.158
equal to or greater than 34	.626

Estimated Recharge

The coefficients (table C12) calculated with equation 1 were used to compute estimated ground-water recharge from each precipitation zone in each valley of the study area (table C19 at the end of this chapter). The results are summarized and compared with ground-water recharge estimated by the reconnaissance studies in table C13. The present study estimates a little more than twice as much recharge from precipitation as was estimated by these earlier studies. The largest percentage changes in estimated recharge were in Antelope, Butte, Goshute, Independence, and Long Valleys, all with present estimates more than 300 percent of the reconnaissance study estimates. Estimates of recharge increased between 200 and 300 percent in Clover, Jakes, Little Smoky, Newark, and Ruby Valleys. Increases in estimated recharge to Railroad, Spring, Steptoe, and Tippett Valleys were less than 200 percent greater than the reconnaissance estimates. Estimated recharge to Little Fish Lake Valley and Hot Creek Valley was about 10 to 20 percent less than that estimated by the reconnaissance study for these valleys.

ESTIMATED GROUND-WATER BUDGETS AND REGIONAL FLOW

Estimated ground-water evapotranspiration for each valley of the study area given in table C7 and the estimated recharge from precipitation given in table C13 were used to develop ground-water budgets for the eastern Nevada study area. The estimates of ground-water evapotranspiration apply specifically and completely to the valley for which the estimate was made. The estimated ground-water recharge applies to the topographic basin of a given valley which is not necessarily coincident with the hydrologic basin. Additionally, the estimated recharge is a bulk estimate that only indirectly implies where the recharge occurs

Table C13. Annual ground-water recharge from precipitation estimated by reconnaissance studies and by present study for valleys of eastern Nevada study area

Valley	Annual ground-water recharge (acre-feet)	
	Reconnaissance study	This study [1]
Antelope Valley	4,700	17,000
Butte Valley	19,000	69,000
Clover Valley	20,700	59,000
Goshute Valley	[2]10,400	41,000
Hot Creek Valley	7,000	5,800
Independence Valley	9,300	50,000
Jakes Valley	17,000	38,500
Little Fish Lake Valley	11,000	9,700
Little Smoky Valley	5,400	13,000
Long Valley	10,300	48,000
Newark Valley	17,500	49,000
Railroad Valley (northern part)	46,000	61,000
Ruby Valley	68,000	146,000
Spring Valley	75,000	104,000
Steptoe Valley	85,000	132,000
Tippett Valley	6,900	12,500
Total	[1]413,000	855,000

[1] Numbers are rounded to the nearest 100 acre-ft for values less than 10,000 acre-ft, to the nearest 500 acre-ft for values equal to or greater than 10,000 acre-ft and less than 100,000 acre-ft, and to the nearest 1,000 acre-ft for values equal to or greater than 100,000 acre-ft.

[2] Includes recharge estimate for northern Antelope Valley.

within the basin, the assumption being that the most recharge occurs in that part of the valley where the most precipitation occurs. Any excess of recharge from precipitation over ground-water evapotranspiration in a given topographic basin is assumed to be discharged from that valley as interbasin flow to adjacent valleys if interbasin flow was proposed by earlier studies or can be supported by available geologic and hydrologic conditions, or represents recharge directly to part of a hydrologic basin that is not coincident with the topographic basin. Similarly, any deficiency in recharge from precipitation is assumed to be compensated by interbasin flow into the valley, or by modifying the topographic basin boundary to include the area of the hydrologic basin to which the recharge occurs. Consequently, any errors in the estimates of ground-water evapotranspiration and recharge are included in the estimates of interbasin ground-water flow.

Harrill and Prudic (1998) identified 287 hydrographic areas in the Great Basin, some of which are divided into from 2 to 4 subareas. These areas and subareas are delineated by topography around and within the valleys for which the areas and subareas were defined; that is, topographic basins and hydrologic areas were assumed to be coincident. Regional ground-water flow systems in the Great Basin have been delineated by Harrill and others (1988). A simulation analysis of regional ground-water flow in the carbonate-rock province of the central and eastern Great Basin, including the present study area, delineated subregional flow-system boundaries on the basis of estimated horizontal ground-water flow and was not constrained by topographic and hydrographic boundaries (Prudic and others, 1995, p. D53). These flow systems are similar, but not identical, to those proposed by Harrill and others (1988).

The results of the present study are consistent with, and tend to corroborate, most of the boundaries previously defined by Harrill and others (1988). To the extent possible, previously delineated regional and subregional flow system names and boundaries of Harrill and others (1988) are retained in the present analysis. Where warranted by the results of the present study, however, flow-system boundaries have been revised. The results of the present study suggest the valleys of the study area may be grouped into the following regional flow systems (pl. 4):

- Newark Valley system –
  - Newark Valley
  - Little Smoky Valley
    - Northern part
    - Central part
- Railroad Valley system –
  - Little Fish Lake Valley
  - Little Smoky Valley
    - Southern part
  - Hot Creek Valley
  - Railroad Valley
    - Northern part
- Independence Valley system –
  - Clover Valley
  - Independence Valley
  - Butte Valley
- Ruby Valley system –
  - Ruby Valley
- Colorado system –
  - Long Valley
  - Jakes Valley

- Goshute Valley system –
  - Goshute Valley
  - Steptoe Valley
- Great Salt Lake Desert system –
  - Spring Valley
  - Tippett Valley
  - Antelope Valley

The Newark and Railroad Valley ground-water flow systems are the same as delineated by Harrill and others (1988). Northern Butte Valley was included in the Ruby Valley system by Harrill and others (1988), but is included in the Independence Valley system in the present study. Southern Butte Valley was included in the Goshute Valley system by Harrill and others (1988), but is included in the Independence Valley system in the present study, although it is possible that the southern one-third of Butte Valley is an isolated subarea that is not connected hydrologically to any adjacent area.

Newark Valley System

The Newark Valley flow system includes Newark Valley and the northern and central parts of Little Smoky Valley to the southwest (pl. 4) and covers the same area proposed by Harrill and others (1988). Annual ground-water discharge by evapotranspiration from Newark Valley was estimated to be about 60,500 acre-ft/yr (tables C7 and C15), which includes from 5,000 to 9,000 acre-ft/yr of evapotranspiration associated with spring discharge in the area of Warm Springs Ranch in northeastern Newark Valley, as much as 6,000 acre-ft/yr from the area south of Buck Mountain, and about 1,500 acre-ft/yr of evapotranspiration from the Fish Creek area in southwestern Newark Valley. Ground-water recharge from precipitation on Newark Valley was estimated to be about 49,000 acre-ft/yr (tables C13 and C15). This suggests an imbalance of about 11,500 acre-ft/yr. The estimated ground-water evapotranspiration and estimated recharge are substantially greater than the 16,000 acre-ft/yr for each estimated by Eakin (1960).

The distribution of ground-water evapotranspiration in Newark Valley, as indicated by the distribution of plant cover shown on plates 2 and 3, suggests sources of recharge, other than precipitation, to the valley. Ground-water evapotranspiration along the northwestern part of the valley obviously is supported by recharge in the Diamond Mountains (pl. 1), while that along the east side of Newark Lake appears to be derived from recharge to Buck Mountain (pl. 1). The estimated ground-water evapotranspiration of 5,000-9,000 acre-ft/yr in the northeast end of the valley, near Warm Springs Ranch (pl. 1), probably is supported by interbasin flow from Long Valley to the east, although some interbasin flow into northern Newark Valley may come from Huntington Valley to the north. The mountains immediately east of Warm Springs Ranch are underlain by Devonian limestones of the Nevada and Devils Gate formations (Hose and Blake, 1976), which were described as having significant transmissivity by Dudley (1967) in the southern Ruby Mountains 20 mi to the north. These formations are part of the lower carbonate aquifer of Winograd and Thordarson (1975) in southern Nevada (fig. C2). An estimated hydraulic gradient of about 200 ft over a distance of about 10 mi (Thomas and others, 1986) and an estimated effective width of ground-water flow of about 25,000 ft underlain by the limestones requires a transmissivity of about 6,000 ft^2/d to yield a flow of 5,000 acre-ft/yr. This suggests a hydraulic conductivity of 6-30 ft/d, well within the range reported by Winograd and Thordarson (1975). The possibility of inflow from Huntington Valley is suggested by sparse water-level data. The presence of a large gravity low beneath the north end of Newark Valley and the south end of Huntington Valley (Ponce, 1992) suggests the possibility of thick basin-fill deposits beneath the topographic divide between the two valleys.

Much of the ground-water evapotranspiration from the phreatophyte area in east-central Newark Valley, south of Buck Mountain, represents an additional estimated 5,000 acre-ft/yr of interbasin flow from Long Valley. The low mountains east of this part of Newark Valley are underlain by the Riepe Spring Limestone and Arcturas Formation, of Pennsylvanian-Permian age and generally equivalent to the upper carbonate aquifer of Winograd and Thordarson (1975) in southern Nevada (fig. C2), and by Tertiary volcanic rocks and small areas of alluvium (Hose and Blake, 1976). An estimated hydraulic gradient of about 200 ft over a distance of about 8 mi (Thomas and others, 1986) and an effective width of as much as 60,000 ft through which ground water may flow requires an average transmissivity of only about 2,600 ft^2/d. This suggests an average hydraulic conductivity of about 2-26 ft/d, depending on aquifer thickness. These values are well within the range of hydraulic conductivities reported by Winograd and Thordarson (1975) and Plume (1996) for the rock types involved. The total interbasin flow from the Long Valley topographic basin to Newark Valley was estimated to be about 10,000 acre-ft/yr (table C14, pl. 4).

The estimated 1,500 acre-ft/yr of evapotranspiration from the Fish Creek area is believed to be supported by interbasin flow beneath the Fish Creek drainage from Little Smoky Valley (pl. 1). Any water that occurs in Newark Lake is derived from surface water runoff and is not part of the ground-water budget. The total estimated interbasin flow to Newark Valley from Long and Little Smoky Valleys satisfies the imbalance in Newark Valley.

Annual ground-water discharge by evapotranspiration from Little Smoky Valley was estimated to be about 6,000 acre-ft/yr (tables C7 and C15), all from the northern part of the valley, as compared to 1,900 acre-ft/yr estimated by Rush and Everett (1966). Annual recharge from precipitation was estimated to be about 13,000 acre-ft/yr (tables C13 and C15); Rush and Everett (1966) estimated 5,400 acre-ft/yr. Of the 7,000 acre-ft/yr of excess recharge to the entire valley (table C14), 1,500 acre-ft/yr was estimated to move out of the basin as interbasin flow to Newark Valley beneath the Fish Creek drainage. The remaining 5,500 acre-ft/yr may leave the southern part of the Valley as interbasin flow to the Lockes area of Railroad Valley, including about 2,500 acre-ft/yr as spring discharge near Lockes.

Railroad Valley System

The Railroad Valley flow system includes Little Fish Lake Valley, the southern part of Little Smoky Valley, Hot Creek Valley, and the northern part of Railroad Valley (pl. 4), and covers the same area proposed by Harrill and others (1988). Estimated recharge to and ground-water evapotranspiration from Little Fish Lake Valley, including about 600 acre-ft/yr of evaporation from Little Fish Lake, are balanced at about 9,700 acre-ft/yr (tables C7, C13, and C15). Rush and Everett (1966) estimated recharge of 11,000 acre-ft/yr and discharge of 10,000 acre-ft/yr for Little Fish Lake Valley. Annual recharge to Hot Creek Valley was estimated by the present study to be about 5,800 acre-ft/yr (tables C13 and C15), slightly less than the 7,000 acre-ft/yr estimated by Rush and Everett (1966). Ground-water discharge by evapotranspiration in Hot Creek Valley was estimated by the present study to be about 5,000 acre-ft/yr (tables C7 and C15), 600 acre-ft/yr more than was estimated by Rush and Everett (1966). The balance of 800 acre-ft/yr is believed to discharge to northern Railroad Valley from the Twin Springs Ranch area (pl. 1) in southeastern Hot Creek Valley (pl. 4, table C14). Ground-water discharge by evapotranspiration from northern Railroad Valley was estimated by the present study to be 85,000 acre-ft/yr (tables C7 and C15), which includes about 50 acre-ft/yr of evaporation from open water estimated to be supplied by ground water. Van Denburgh and Rush (1974) estimated ground-water evapotranspiration from northern Railroad Valley to be about 80,000 acre-ft/yr. Ground-water recharge from precipitation is estimated to be about 61,000 acre-ft/yr (tables C13 and C15), about 30 percent more than the 46,000 acre-ft/yr estimated by Eakin and others (1951b) and Van Denburgh and Rush (1974).

The northern part of Railroad Valley receives additional recharge in the form of interbasin flow (table C14, pl. 4): about 800 acre-ft/yr from the Twin Springs Ranch area of Hot Creek Valley; about 5,500 acre-ft/yr from the southern part of Little Smoky Valley that is discharged near Lockes in Railroad Valley; about 13,000 acre-ft/yr that is discharged near Duckwater in northwestern Railroad Valley (Chapter B of this report); and about 4,000 acre-ft/yr of northward flow from the southern part of Railroad Valley (Van Denburgh and Rush, 1974, p. 25).

The spring discharge at Duckwater is believed to be derived from deep interbasin ground-water flow from Long Valley to the northeast. Van Denburgh and Rush (1974) suggested the spring discharge at Duckwater entered Railroad Valley from adjacent, but as yet unidentified, valleys. Mifflin (1968) provided geochemical evidence that water discharging at Duckwater springs is related to regional ground-water flow systems and therefore originates from outside Railroad Valley. Harrill and others (1988) suggest interbasin flow southward from Newark Valley toward the Duckwater area. Prudic and others (1995) suggested that deep interbasin ground-water flow occurs along a zone of high transmissivity extending from northern Long Valley through eastern Newark Valley to the Duckwater area of Railroad Valley and perhaps continuing south. The present study suggests that Long Valley is the ultimate source of spring discharge in the Duckwater area because of the excess of recharge over ground-water evapotranspiration in Long Valley, although this interbasin flow probably moves beneath Newark Valley. Previous studies were unable to arrive at any conclusions regarding the source of this discharge because the reconnaissance ground-water budgets for Newark and Long Valleys were substantially balanced with no excess recharge available to allocate to interbasin flow.

Table C14. Estimated distribution of interbasin ground-water flow for valleys in eastern Nevada study area

[All values are in acre-feet per year]

Valley	Interbasin flow from	Interbasin flow	Total	Interbasin flow to	Interbasin flow	Total
Antelope Valley	Goshute Valley	500		Great Salt Lake Desert	13,500	
			500			13,500
Butte Valley				Clover Valley	22,500	
				Ruby Valley	2,000	
						24,500
Clover Valley	Butte Valley	22,500				
	Independence Valley	3,000				
			25,500			
Goshute Valley	Steptoe Valley	4,000		Antelope Valley	500	
				Great Salt Lake Desert	2,000	
			4,000			2,500
Hot Creek Valley				Railroad Valley	800	
				(northern part)		800
Independence Valley				Clover Valley	3,000	3,000
Jakes Valley	Long Valley	14,000		White River Valley	51,200	
				Railroad Valley		
				(northern part)	700	
			14,000			51,900
Little Smoky Valley				Newark Valley	1,500	
				Railroad Valley		
				(northern part)	5,500	
						7,000
Long Valley				Newark Valley	10,000	
				Jakes Valley	14,000	
				Railroad Valley		
				(northern part)	13,000	
						37,000
Newark Valley	Long Valley	10,000				
	Little Smoky Valley	1,500				
			11,500			
Railroad Valley (northern part)	Long Valley	13,000				
	Little Smoky Valley	5,500				
	Hot Creek Valley	800				
	Railroad Valley					
	(southern part)	4,000				
	Jakes Valley	700				
			24,000			
Ruby Valley	Huntington Valley	19,000				
	Butte Valley	2,000				
			21,000			
Spring Valley				Hamlin Valley	10,000	
				Snake Valley	4,000	
						14,000
Steptoe Valley				Goshute Valley		4,000
Tippett Valley				Great Salt Lake Desert	6,000	
				Snake Valley	3,600	
						9,600

Table C15. Summary of estimated ground-water budgets for valleys in eastern Nevada study area

Valley	Recharge (acre-feet per year)			Discharge (acre-feet per year)		
	From precipitation	Interbasin flow in	Total	Evapotranspiration	Interbasin flow out	Total
Antelope Valley	17,000	500	17,500	4,000	13,500	17,500
Butte Valley	69,000	0	69,000	44,500	24,500	69,000
Clover Valley	59,000	25,500	84,500	84,500	0	84,500
Goshute Valley	41,000	4,000	45,000	42,500	2,500	45,000
Hot Creek Valley	5,800	0	5,800	5,000	800	5,800
Independence Valley	50,000	0	50,000	47,000	3,000	50,000
Jakes Valley	38,500	14,000	52,500	600	51,900	52,500
Little Fish Lake Valley	9,700	0	9,700	9,700	0	9,700
Little Smoky Valley	13,000	0	13,000	6,000	7,000	13,000
Long Valley	48,000	0	48,000	11,000	37,000	48,000
Newark Valley	49,000	11,500	60,500	60,500	0	60,500
Railroad Valley (northern part)	61,000	24,000	85,000	85,000	0	85,000
Ruby Valley	146,000	21,000	167,000	167,000	0	167,000
Spring Valley	104,000	0	104,000	90,000	14,000	104,000
Steptoe Valley	132,000	0	132,000	128,000	4,000	132,000
Tippett Valley	12,500	0	12,500	2,900	9,600	12,500

An estimated 5,500 acre-ft/yr of interbasin flow from Little Smoky Valley in the Lockes area of Railroad Valley includes about 2,500 acre-ft/yr of discharge from springs near Lockes. In Chapter B of this report, about 9,000 acre-ft/yr of ground-water evapotranspiration is estimated for the Lockes area, much of which may be derived from interbasin flow from Little Smoky Valley. The occurrence of interbasin flow from Little Smoky Valley to Railroad Valley is consistent with previous studies (Rush and Everett, 1966; Van Denburgh and Rush, 1974). The amount of interbasin flow estimated from southern Little Smoky Valley was selected to balance the ground-water budget for Little Smoky Valley.

The interbasin flow discussed above accounts for 23,300 acre-ft/yr, which is 700 acre-ft/yr less than needed to balance the estimated ground-water evapotranspiration of 85,000 acre-ft/yr. The difference may arise because the estimated ground-water evapotranspiration is too large, or because the estimated recharge from precipitation is too small, or because the interbasin flow from Long Valley by way of Newark Valley is too small, or because interbasin flow may enter Railroad Valley from Jakes Valley to the northeast as suggested by Harrill and others (1988), or through a combination of these factors. Interbasin flow of 700 acre-ft/yr from Jakes Valley is assumed by the present study (table C14, pl. 4) until a more detailed investigation of interbasin flow to Railroad Valley is made.

Total interbasin flow to northern Railroad Valley was estimated to be about 24,000 acre-ft/yr (table C14) compared to 7,000 acre-ft/yr estimated by Van Denburgh and Rush (1974). Total estimated ground-water recharge from precipitation and interbasin flow to northern Railroad Valley was estimated to be about 85,000 acre-ft/yr (tables C13 and C15).

Independence Valley System

The Independence Valley flow system includes Independence, Clover, and Butte Valleys. The inclusion of Butte Valley in the Independence Valley flow system differs from the geographic area proposed by Harrill and others (1988); they included the northern Butte Valley hydrographic area, which includes only the northern one-third of Butte Valley, in the Ruby Valley flow system and the southern Butte Valley hydrographic area, which includes the southern two-thirds of Butte Valley, in the Goshute Valley system. Given the interpretation of ground-water flow for this system by the present study, the system might more properly be named the Clover Valley flow system, but the name given to the system by Harrill and others (1988) is retained for consistency.

Eakin and Maxey (1951b) estimated recharge to Independence Valley to be 9,300 acre-ft/yr and discharge to be 9,500 acre-ft/yr. For Clover Valley, they estimated recharge of 20,700 acre-ft/yr and discharge of 19,000 acre-ft/yr. Glancy (1968) estimated ground-water recharge to Butte Valley of 19,000 acre-ft/yr, discharge of 19,900 acre-ft/yr, and estimated 800 acre-ft/yr of interbasin flow from Butte Valley to Ruby Valley. Interbasin flow from Butte Valley to Clover Valley was not suggested by the early investigations, although Harrill and others (1988) suggested minor interbasin flow from Clover Valley to Independence Valley and 1,000 acre-ft/yr of interbasin flow from Butte Valley to Ruby Valley. Prudic and others (1995, p. D75) suggested interbasin flow to Clover Valley of 9,000 acre-ft/yr from the upper Humboldt River region north and northwest of Clover Valley.

Recharge from precipitation to Clover Valley was estimated by the present study to be about 59,000 acre-ft/yr (tables C13 and C15) and ground-water evapotranspiration from Clover Valley was estimated to be about 84,500 acre-ft/yr (tables C7 and C15), leaving an imbalance of about 25,500 acre-ft/yr. Recharge from precipitation to Independence Valley was estimated to be about 50,000 acre-ft/yr (tables C13 and C15). Estimated ground-water evapotranspiration was estimated to be 47,000 acre-ft/yr (tables C7 and C15). The remaining 3,000 acre-ft/yr was assumed to move as interbasin flow to Clover Valley to the west (table C14, pl. 4). Ground-water level data are not available to determine the direction of ground-water flow, if any, between Independence and Clover Valleys. Annual recharge from precipitation to Butte Valley, south of Clover Valley, was estimated to be about 69,000 acre-ft/yr (tables C13 and C15); ground-water evapotranspiration from Butte Valley was estimated to be 44,500 acre-ft/yr (tables C7 and C15). Of the remaining 24,500 acre-ft/yr, 22,500 acre-ft/yr are assumed to move as interbasin flow northward into Clover Valley and 2,000 acre-ft/yr are assumed to move as interbasin flow to Ruby Valley (table C14, pl. 4).

The topographic divide between Clover and Butte Valleys is low in altitude and underlain by alluvium to unknown depths. Geophysical evidence (Ponce, 1992; Ponce and others, 1996) suggests a continuous series of deep bedrock basins beneath the valley floor extending from about the middle of the Cherry Creek Range, on the east side of Butte Valley, northward into central Clover Valley. Although the valleys are distinct topographically, Clover Valley and the northern two thirds of Butte Valley may constitute a single hydrologic basin. The southern third of Butte Valley appears to consist of two small deep bedrock basins (Ponce, 1992), one west of the southern Cherry Creek Range and the other south of the range. The southern one-third of Butte Valley may be an isolated, hydrologically closed basin or may be hydrologically associated with Steptoe Valley as suggested by Harrill and others (1988). For the analysis herein, all of Butte Valley is assumed to be part of the Independence Valley flow system.

Ground-water levels in northern Butte Valley, southern Clover Valley, and eastern Ruby Valley suggest the possibility of interbasin flow from Butte Valley to adjacent areas of Clover and Ruby Valleys (Thomas and others, 1986). The steepest hydraulic gradient, however, is between Butte Valley and Clover Valley suggesting greater interbasin flow between these two valleys. The hydraulic gradient from the Christiansen well (Township 30 North, Range 62 East, section 33) just north of West Buttes in northern Butte Valley to the Spruce well (Township 31 North, Range 62 East, section 3) in southern Clover Valley is about 290 ft in about 12 mi (Thomas and others, 1986). The width of alluvial fill near the topographic divide between the valleys is about 27,000 ft at land surface. However, bedrock of the bounding mountain ranges may not limit flow northward because limestones of the Permian Pequop Formation outcrop on the west side of the Clover Valley–Butte Valley divide area (Coats, 1987), while carbonate rocks of the Ordovician Pogonip Group and of the Pennsylvanian-Permian (undivided) Riepe Spring Limestone and Rib Hill Formation (Coats, 1987) underlie Spruce mountain on the east side of the divide area. These formations are included in, or are equivalent to, the lower and upper carbonate aquifers of Winograd and Thordarson (1975) in southern Nevada (fig. C2). Assuming a section width of 22,000 ft and the hydraulic gradient given above, a transmissivity of about 27,000 ft^2/d would be required to allow ground-water flow of 22,500 acre-ft/yr from Butte Valley to Clover Valley. This transmissivity is within the range of transmissivity reported for the lower and upper carbonate aquifers in southern Nevada (Winograd and Thordarson, 1975, p. C30, C34). The average hydraulic conductivity would have to be from about 30 to 60 ft/d, well within the range of hydraulic conductivity for carbonate-rock aquifers and alluvial valley-fill materials reported by Plume (1996).

Glancy (1968) reported a gradient of 40 ft over 5.5 mi between northern Butte Valley and Ruby Valley to the west. Thomas and others (1986), using water-level altitudes based on more current field mapping indicate a gradient of 25 ft over 5.4 mi. Using this gradient, a flow section width of 10,560 ft, as estimated by Glancy (1968) and a transmissivity of 27,000 ft^2/d suggests interbasin flow of as much as 2,000 acre-ft/yr from Butte Valley to Ruby Valley.

About 9,000 acres of northwestern Clover Valley along the eastern base of the East Humboldt Range are covered by native meadows and pasture. The meadows appear to be supported by shallow ground water derived from locally large surface-water discharge from watersheds in the East Humboldt Range, which are underlain by granitic and metamorphic rocks of uncertain age. This surface discharge does not have an opportunity to infiltrate into the subsurface until it leaves the bedrock areas of the watersheds in the mountain block. Ground-water evapotranspiration from these areas, areas which were not included in the reconnaissance estimates of ground-water evapotranspiration (Eakin and Maxey, 1951b), was estimated to be about 23,000 acre-ft/yr.

Ruby Valley System

The Ruby Valley system includes only Ruby Valley in the present study and differs from the geographic area delineated by Harrill and others (1988) by excluding the northern part of Butte Valley, although a small amount of ground water flows from Butte Valley into Ruby Valley. Ground-water recharge from precipitation to Ruby Valley was estimated to be about 146,000 acre-ft/yr (tables C13 and C15). Eakin and Maxey (1951a) estimated 68,000 acre-ft/yr of ground-water recharge to Ruby Valley.

Ground-water evapotranspiration was estimated to be 167,000 acre-ft/yr (tables C7 and C15). Some of Ruby Lake and most, if not all, of Franklin Lake, when it is present, is derived from surface-water runoff. Ruby Lake may cover from as little as 1,000 acres to more than 3,500 acres. Franklin Lake is dry in many years, but may cover 30,000 to 45,000 acres in wet years. The two Lakes covered more than 49,000 acres in 1985, but by 1989 had been reduced to just over 1,000 acres, all of it in Ruby Lake. Eakin and Maxey (1951a) estimated ground-water discharge of 67,600 acre-ft/yr.

About 23,500 acres of the west side of northern Ruby Valley along the eastern base of the Ruby Mountains are covered by native meadows and pastures, similar to those found in Clover Valley. Hydrologic conditions in northern Ruby Valley are similar to those in Clover Valley; locally large surface discharge occurs from the watersheds of the northern Ruby Mountains, which are underlain by Tertiary and Cretaceous granitic rocks and lower Paleozoic metamorphic rocks. Ground-water evapotranspiration from these areas was estimated to be about 47,500 acre-ft/yr. These areas were not included in areas of ground-water evapotranspiration by the reconnaissance study of Eakin and Maxey (1951a).

The earlier discussion of the marsh area of southern Ruby Valley concluded that about 22,000 acre-ft/yr of ground water is evapotranspired from the 14,600 acres of marshland. This is in addition to the estimated ground-water evapotranspiration of 145,000 acre-ft/yr from non-marsh vegetation in Ruby Valley, for a total estimated ground-water evapotranspiration of 167,000 acre-ft/yr (table C17). This exceeds the estimated recharge of 146,000 acre-ft/yr from precipitation by 21,000 acre-ft/yr. Ground-water levels in southern Ruby Valley are higher than those in the Franklin Lake area just north of Ruby Lake. Northern Ruby Valley is an unlikely source of any ground water with which to support the marshes of southern Ruby Valley. The most likely source is interbasin flow of recharge from precipitation on the west side of the southern Ruby Mountains that moves through the thick carbonate-rock sequence between Harrison Pass and Overland Pass (pl. 1).

Eakin and Maxey (1951a) did not suggest interbasin ground-water flow into southern Ruby Valley, but did indicate that discharge from springs along the southwestern part of the valley may be substantially greater than the amount used in their estimates (Eakin and Maxey, 1951a, p. 83); the discharge estimate they used is not clear. Dudley (1967) estimated between 9,000 and 15,000 acre-ft/yr of interbasin flow from the west slope of the southern Ruby Mountains to Ruby Valley along the western margin of the marshes of Ruby Lake based on an analysis of the hydrogeology of the Ruby Mountains and electric-analog modeling of ground-water flow. Harrill and others (1988) suggested interbasin flow of 10,000 acre-ft/yr from Huntington Valley, to the west, into southern Ruby Valley. Prudic and others (1995, p. D75) estimated interbasin flow of 16,000 acre-ft/yr from Huntington Valley.

The southern Ruby Mountains between Harrison Pass and Overland Pass (pl. 1), which border southern Ruby Valley to the west of Ruby Lake, is an asymmetrical fault-block mountain with a steep east-facing slope and gentler western slopes. This part of the Ruby Mountains is underlain by a thick sequence of eastward-dipping, mostly carbonate-rock formations (fig. C2) ranging in age from Cambrian to Pennsylvanian (Dudley, 1967; Coats, 1987). At the base of the western slope of the southern Ruby Mountains is a 4,000-ft sequence of middle Cambrian limestones to which Dudley (1967) assigned a medium to locally high permeability. Above these limestones are limestones of the 3,000-ft thick Pogonip Group, described by Dudley (1967) as the most important aquifer in the range because of its secondary porosity and permeability and because of its structural position low on the western slopes. Above the Pogonip is a 3,000-ft thick sequence of Silurian and Devonian dolomites of moderate permeability (Dudley, 1967). Finally, above these dolomites are 1,700 ft of Devonian limestones of the Nevada Formation and Devils Gate limestone, most of which are highly permeable (Dudley, 1967). These formations are the lower carbonate aquifer of Winograd and Thordarson (1975) in southern Nevada (fig. C2). Capping the range in some localities are limestones of Pennsylvanian and Permian age that are hydrologically unimportant because of their topographic position (Dudley, 1967).

The contour line representing 20 inches of precipitation on the PRISM precipitation map occurs just above the base of the middle Cambrian limestones on the western slopes of the southern Ruby Mountains. Applying the recharge percentages to precipitation of 20 inches and greater falling on the western slopes of the mountains yields recharge of about 22,500 acre-ft/yr. This recharge, if most or all of it moves down-dip through the thick, eastward-dipping limestones and dolomites of the southern Ruby Mountains, is sufficient to provide the additional recharge needed to balance the ground-water budget for Ruby Valley. The present study suggests that as much as 19,000 acre-ft/yr of recharge to Ruby Valley may be through interbasin flow from the west slope of the southern Ruby Mountains and 2,000 acre-ft/yr from northern Butte Valley (pl. 4, table C14).

Colorado System

The Colorado flow system of Eakin (1966) and later Harrill and others (1988) extends far south of the study area. Valleys in the study area included in this system are the northernmost valleys of the Colorado system—Long and Jakes Valleys.

Recharge from precipitation to Long Valley was estimated to be about 48,000 acre-ft/yr (tables C13 and C15) and ground-water evapotranspiration was estimated to be about 11,000 acre-ft/yr (tables C7 and C15). About 13,000 acre-ft/yr is believed to move as deep interbasin flow, perhaps through Newark Valley, to the springs at Duckwater in Railroad Valley (pl. 4, table C14). An estimated 10,000 acre-ft/yr of additional interbasin flow from Long Valley sustains phreatophyte evapotranspiration in northeastern and east-central Newark Valley, described above (pl. 4, table C14). The remaining 14,000 acre-ft/yr of recharge is believed to move as deep interbasin flow into and through Jakes Valley to the south (pl. 4, table C14). Significant deep interbasin flow out of Long Valley has been proposed by a number of studies, including those by Eakin (1961, 1966), Rush and others (1971), Harrill and others (1988), and Prudic and others (1995). Early studies estimated recharge to Long Valley of 10,300 acre-ft/yr, evapotranspiration of 2,200 acre-ft/yr, and interbasin flow out of the valley of 8,100 acre-ft/yr (Eakin, 1961, 1966).

Annual recharge from precipitation to Jakes Valley was estimated by the present study to be about 38,500 acre-ft/yr (tables C13 and C15); ground-water evapotranspiration, along Illipah Creek west of the main basin of Jakes Valley, was estimated to be about 600 acre-ft/yr (tables C7 and C15). The balance of 37,900 acre-ft/yr together with the 14,000 acre-ft/yr of interbasin flow from Long Valley leaves Jakes Valley as interbasin flow to upper White River Valley to the south (table C14), although 700 acre-ft/yr may move as interbasin flow to Railroad Valley. The total interbasin flow into White River Valley from Long and Jakes Valleys is estimated to be about 51,000 acre-ft/yr. Eakin (1966) estimated recharge to Jakes Valley at 17,000 acre-ft/yr and did not quantify ground-water evapotranspiration.

The estimated ground-water recharge and discharge for Jakes Valley are problematic. Jakes Valley is one of the high valleys of eastern Nevada, with the valley floor at or above an altitude of 6,300 ft. Consequently, estimated annual precipitation on the valley floor is equal to or greater than 12 inches. An estimated 26,000 acre-ft/yr of recharge would be derived from precipitation on the valley floor. With ground-water levels estimated to be as much as 400 ft below the playa of Jakes Valley (Eakin, 1966), whether or not all this estimated recharge reaches the ground-water system is questionable. Gravity data suggest an alignment of deep bedrock basins southward from Long Valley, through Jakes Valley, into and down White River Valley (Ponce, 1992; Snyder and others, 1984). Extensive areas of spring discharge exist in the upper White River Valley, but to develop a new ground-water budget, the White River Valley would have to be studied to determine if the magnitude of interbasin flow from Long and Jakes Valleys suggested above can be substantiated.

Goshute Valley System

The Goshute Valley flow system includes Goshute Valley, Steptoe Valley, and perhaps the southern one-third of Butte Valley in the present study. Harrill and others (1988) included the southern two-thirds of Butte Valley in the Goshute Valley system. Ground-water recharge from precipitation to Goshute Valley was estimated to be about 41,000 acre-ft/yr and ground-water evapotranspiration was estimated to be 42,500 acre-ft/yr (tables C7, C13, and C15). This compares to recharge of 10,400 acre-ft/yr and evapotranspiration of 10,075 acre-ft/yr estimated by Eakin and others (1951a) for Goshute-Antelope Valley, an area larger than the Goshute Valley as defined for the present study. Ground-water recharge from precipitation to Steptoe Valley was estimated to be about 132,000 acre-ft/yr. Ground-water evapotranspiration from the valley was estimated to be about 128,000 acre-ft/yr (tables C7, C11, and C15). This compares to an estimated recharge of 85,000 acre-ft/yr and discharge of 71,000 acre-ft/yr estimated by Eakin and others (1967).

Harrill and others (1988) have suggested about 2,000 acre-ft/yr of interbasin flow from Goshute Valley to the Great Salt Lake Desert to the east and minor interbasin flow into Antelope Valley to the southeast. Eakin and others (1967) suggested about 1,000 acre-ft/yr of interbasin flow from Steptoe Valley north of Currie into Goshute Valley in the Nelson Creek area (pl. 1). Allowing 4,000 acre-ft/yr of interbasin flow from Steptoe Valley, north of Currie, into Goshute Valley, balances the ground-water budget for Steptoe Valley (pl. 4) and provides sufficient additional recharge to Goshute Valley to balance the budget there also (pl. 4; tables C14 and C15).

Great Salt Lake Desert System

Antelope, Tippett, and Spring Valleys in eastern Nevada are part of the Great Salt Lake Desert flow system of Harrill and others (1988), which includes all or parts of 13 more valleys in western Utah. Recharge from precipitation to Antelope Valley was estimated by the present study to be about 17,000 acre-ft/yr and ground-water evapotranspiration was estimated to be about 4,000 acre-ft/yr (tables C7, C13, and C15). Recharge to Tippett Valley was estimated to be about 12,500 acre-ft/yr and ground-water evapotranspiration was estimated to be about 2,900 acre-ft/yr (tables C7, C13, and C15). The large difference between estimated recharge and estimated ground-water evapotranspiration for Antelope and Tippett Valleys may, in part, result from errors in both estimates, but the difference is so great that recharge to these two valleys probably exceeds estimated ground-water evapotranspiration by a considerable amount. Significant interbasin ground-water flow from these valleys has been suggested by an earlier study. Harrill (1971) proposed interbasin flow of 300 acre-ft/yr from Goshute Valley to Antelope Valley and 4,900 acre-ft/yr from Antelope Valley to the Great Salt Lake Desert to the east. He also proposed 5,000 acre-ft/yr of interbasin flow from Tippett Valley to the Great Salt Lake Desert and 2,000 acre-ft/yr from Tippett Valley to Spring Valley. Harrill and others (1988) suggested minor interbasin flow into Antelope Valley from Goshute Valley; 8,000 acre-ft/yr of interbasin flow from Antelope Valley to the Great Salt Lake Desert; 2,000 acre-ft/yr of interbasin flow from Tippett Valley to the Great Salt Lake Desert; 3,000 acre-ft/yr from Tippett Valley to Antelope Valley; and 2,000 acre-ft/yr from Tippett Valley to Spring Valley.

The present study suggests that all recharge to Antelope and Tippett Valleys in excess of ground-water evapotranspiration discharges ultimately to the Great Salt Lake Desert. About 13,500 acre-ft/yr of recharge to Antelope Valley are estimated to leave Antelope Valley as interbasin flow to the Great Salt Lake Desert to the east (pl. 4, table C14) and about 9,600 acre-ft/yr are estimated to leave Tippett Valley as interbasin flow (pl. 4, tables C14 and C15), with about 6,000 acre-ft/yr flowing through southern Antelope Valley and ultimately to the Great Salt Lake Desert. The remaining 3,600 acre-ft/yr of interbasin flow from Tippett Valley is to the Great Salt Lake Desert in the area of Salt Lake Marsh. Much of the recharge to Antelope Valley occurs in the Goshute Mountains north of Alternate U.S. Highway 93 (pl. 1). The mountains in this area are underlain mainly by Devonian carbonate rocks, part of the lower carbonate aquifer of Winograd and Thordarson (1975). Much of the recharge to Tippett Valley occurs in the Antelope Range, which bounds Tippett Valley on the west. These mountains, in Tippett Valley, are underlain by lower Paleozoic carbonates, including the Pogonip limestone and Devonian carbonate rocks of the lower carbonate aquifer of Winograd and Thordarson (1975), and carbonate rocks of the Permian Arcturus Formation, part of the upper carbonate aquifer of Winograd and Thordarson (1975). A smaller area of recharge occurs in southeastern Tippet Valley in the Kern Mountains, which are comprised largely of Tertiary volcanic rocks, but are flanked north and south by carbonate rocks of Devonian and Permian age.

Water-level data are sparse in these two valleys (Harrill, 1971; Thomas and others, 1986). The data for Tippett Valley are ambiguous, but suggest much of the available recharge from the Antelope Range may flow northward through southern Antelope Valley or northeastward into Deep Creek Valley (pl. 1). About 6,000 acre-ft/yr of ground water from Tippett Valley are believed to join 13,000 acre-ft/yr of ground water from Antelope Valley and discharge at Big Salt Spring and Little Salt Spring near Blue Lakes (pls. 1 and 4) east of the Goshute Mountains that border Antelope Valley on the northeast. Gates and Kruer (1981, p. 18) suggested that all the measured 19,000 acre-ft/yr of discharge at these springs was derived from recharge to topographic basins in Nevada. Recharge to the Kern Mountains in southeastern Tippett Valley probably moves down Pleasant Valley, north of the Kern Mountains, and through the carbonate rocks and alluvium south of the mountains to Snake Valley (pls. 1 and 4). Altogether, about 3,600 acre-ft/yr of recharge originating in southeastern Tippett Valley may move as interbasin flow to Snake Valley.

Recharge from precipitation to Spring Valley was estimated to be about 104,000 acre-ft/yr. Ground-water evapotranspiration from Spring Valley was an estimated 90,000 acre-ft/yr (tables C7, C13, and C15). The difference of 14,000 acre-ft/yr may be the result of an underestimation of discharge by evapotranspiration of ground-water, an overestimation of recharge, or a combination of errors in both estimates. However, much of the excess recharge is believed to leave the valley as interbasin flow to the east (pl. 4, tables C14 and C15). Rush and Kazmi (1965) estimated about 4,000 acre-ft/yr of subsurface outflow of ground water from Spring Valley to Hamlin Valley through the carbonate rocks of the southern Snake Range. Field observation indicates the southern end of the phreatophyte area in southern Spring Valley is about coincident with the intersection of southeastward-dipping massive Upper Cambrian limestones and the valley floor. These limestones are one of the major water-bearing units of the lower carbonate aquifer (fig. C2) of Winograd and Thordarson (1975). Rush and Kazmi (1965), while suggesting that the subsurface flow is principally through the carbonate rocks of the southern Snake Range, use an estimated transmissivity for the alluvium of about 6,700 ft^2/d in calculating the flow. This implied a hydraulic conductivity as small as about 2 ft/d, considerably lower than the mean and median values of 80 and 6 ft/d for the hydraulic conductivity of the lower carbonate aquifer reported by Winograd and Thordarson (1975). The estimated interbasin flow of Rush and Kazmi (1965), therefore, can be increased easily by a factor of two or three to an estimated 8,000 to 12,000 acre-ft/yr.

Another area where interbasin flow from Spring Valley may occur is in the area between the Kern Mountains and the northern end of the Snake Range (pl. 1). This low-altitude area is underlain by an east-west trending series of gravity lows (Ponce, 1992) and by sedimentary deposits surrounding outcrops of Tertiary volcanic rocks and carbonate rocks of the Permian Arcturus Formation; the Pogonip limestone outcrops on the east side of the Red Hills just north of these alluvial deposits (Hose and Blake, 1976). Water-level data for the area of Spring Valley just to the west of this low-altitude area are sparse and ambiguous (Rush and Kazmi, 1965; Thomas and others, 1986). Interbasin flow may be significant from Spring Valley through the sedimentary and carbonate-rock deposits of this area to a large area of phreatophytes around Salt Marsh Lake near Gandy, Utah. Additional hydrologic and hydrogeologic studies in Snake and Hamlin Valleys and the Great Salt Lake Desert are needed to corroborate the estimates of interbasin flow from Antelope, Tippett, and Spring Valleys.

SUMMARY AND CONCLUSIONS

Ground-water budgets were estimated for 16 contiguous valleys covering nearly 15,000 mi^2 of the central Great Basin in Elko, White Pine, Nye, Eureka, and Lincoln Counties of eastern Nevada. The geologically complex terrain of the study area consists of valleys underlain by unconsolidated deposits of Quaternary age bounded by mountain ranges of folded and faulted rocks ranging in age from Precambrian to late Tertiary. The valleys and bounding mountain ranges generally trend north-south with the exception of the Ruby Mountains and the East Humboldt, Quinn Canyon, and

Grant Ranges, which trend northeast-southwest. The climate of the area ranges from an arid middle-latitude desert on thc valley floors to a subhumid continental climate in the highest mountain ranges. Vegetation varies from desert scrub at the lowest altitudes to subalpine forest and alpine tundra at the highest altitudes.

Early ground-water resources studies in the Nevada part of the Great Basin recognized that most ground water is discharged from the topographically closed valleys by evapotranspiration. Reconnaissance-level studies from the late 1940's to the mid-1970's estimated ground-water budgets for the valleys of the study area. Ground-water evapotranspiration estimated by these studies was based on the results of field studies made before 1930. Ground-water recharge estimated by these studies was based on precipitation data first published in 1936. However, the reconnaissance level ground-water budgets should be reconsidered because of the results of ground-water evapotranspiration field studies since 1980, the availability of more recent precipitation data, and more advanced technologies.

The results of the present study are based on these more recent data, methods, and technologies. Ground-water discharge by evapotranspiration was estimated from field studies of phreatophyte evapotranspiration, the results of which were correlated to plant cover. Estimates of phreatophyte plant cover were determined at regional scales from a vegetation index derived from Landsat data. Ground-water evapotranspiration estimated by the present study totaled about 790,000 acre-ft/yr, and exceeded previous estimates by a little more than two times. A new precipitation map for Nevada, based on the PRISM methodology, was used to estimate the volume of precipitation in each valley of the study area. New estimates of ground-water recharge from precipitation were based on the new estimates of ground-water evapotranspiration and interbasin ground-water flow in 15 of the 16 valleys of the study area. No recharge was assumed for areas with annual precipitation of less than 8 inches. Recharge was estimated to be about 0.8 percent of annual precipitation from 8 to less than 12 inches, about 13.0 percent of precipitation from 12 to less than 16 inches, about 14.4 percent of precipitation from 16 to less than 20 inches, about 15.8 percent of precipitation from 20 to less than 34 inches, and about 62.6 percent of precipitation equal to or greater than 34 inches. Ground-water recharge estimated by the present study totaled about 855,000 acre-ft/yr, a little more than twice the amount estimated by previous studies.

The estimates of ground-water discharge by evapotranspiration and ground-water recharge from precipitation were used to develop revised ground-water budgets for all or parts of seven regional ground-water flow systems in eastern Nevada (table C16). These flow systems include the Newark Valley, Railroad Valley, Independence Valley, Ruby Valley, Colorado, Goshute Valley, and Great Salt Lake Desert flow systems. The estimated ground-water budget for each valley is based on the conventional assumption that hydrologic boundaries are coincident with topographic boundaries. Budgets for individual valleys were balanced by assuming interbasin ground-water flow from or to adjacent valleys within the boundaries of the

Table C16. Summary of estimated ground-water budgets for regional flow systems in eastern Nevada study area and for overall study area

Regional flow system	Recharge (acre-feet per year)			Discharge (acre-feet per year)		
	From precipitation	Interbasin flow in	Total	Evapotranspiration	Interbasin flow out	Total
Newark Valley system	56,500	10,000	66,500	66,500	0	66,500
Railroad Valley system	82,000	17,700	99,700	99,700	0	99,700
Independence Valley system	178,000	0	178,000	176,000	2,000	178,000
Ruby Valley system	146,000	21,000	167,000	167,000	0	167,000
Colorado system	86,500	0	86,500	11,600	74,900	86,500
Goshute Valley system	173,000	0	173,000	170,500	2,500	173,000
Great Salt Lake Desert system	133,500	500	134,000	96,900	37,100	134,000
Overall study area [1]	855,000	[2]23,000	[2]878,000	788,000	[2]90,000	[2]878,000

[1] Numbers are rounded to the nearest 1,000 acre-feet per year.

[2] Total includes only interbasin ground-water flow entering or leaving overall study area (interbasin flow between flow systems within study area is not included).

flow system. Much of the assumed interbasin flow may more properly reflect recharge in areas where hydrologic divides diverge from topographic divides.

Recharge from precipitation to the Newark Valley ground-water flow system, which is comprised of Newark Valley and northern and central Little Smoky Valley, was estimated to be about 56,500 acre-ft/yr. An estimated 10,000 acre-ft/yr of ground water is assumed to enter the Newark Valley system as interbasin flow from Long Valley. Ground-water evapotranspiration from the system was estimated to total about 66,500 acre-ft/yr.

The Railroad Valley system includes Little Fish Lake Valley, the southern part of Little Smoky Valley, Hot Creek Valley, and the northern part of Railroad Valley. Total recharge from precipitation to this system was estimated to be about 82,000 acre-ft/yr. Regional interbasin ground-water flow into the Railroad Valley system from Long Valley, probably through Newark Valley, was estimated to be about 13,000 acre-ft/yr. Another 700 acre-ft/yr may enter Railroad Valley from Jakes Valley and about 4,000 acre-ft/yr may enter the northern part of Railroad Valley from the southern part. Total recharge to the Railroad Valley system was estimated to be about 99,700 acre-ft/yr. Estimated ground-water evapotranspiration from the valleys of the Railroad Valley system also totaled about 99,700 acre-ft/yr.

Recharge from precipitation to the Independence Valley system, which includes Independence, Clover, and Butte Valleys, was estimated to be about 178,000 acre-ft/yr and ground-water evapotranspiration was estimated to be about 176,000 acre-ft/yr; 2,000 acre-ft/yr may move as interbasin flow from northern Butte Valley to Ruby Valley.

The Ruby Valley flow system includes only Ruby Valley. Total recharge from precipitation was estimated to be 146,000 acre-ft/yr; as much as 19,000 acre-ft/yr are assumed to enter Ruby Valley as interbasin flow from precipitation on the west slopes of the southern Ruby Mountains. Another 2,000 acre-ft/yr are assumed to recharge Ruby Valley as interbasin flow from northern Butte Valley. Total ground-water evapotranspiration was estimated to be about 167,000 acre-ft/yr.

Long and Jakes Valleys are the northern-most valleys of the Colorado system, which stretches far south of the study area. Recharge from precipitation to Long and Jakes Valleys was estimated to be about 86,500 acre-ft/yr while estimated ground-water evapotranspiration was about 11,600 acre-ft/yr. About 10,000 acre-ft/yr was estimated to move as interbasin flow from Long Valley to Newark Valley. Another 13,000 acre-ft/yr was estimated to leave Long Valley, possibly flowing through southern Newark Valley, as regional interbasin flow to springs in Railroad Valley near Duckwater. The remaining 14,000 acre-ft/yr of recharge to Long Valley is assumed to move as interbasin flow, together with as much as 37,200 acre-ft/yr of recharge to Jakes Valley, to the White River Valley. About 700 acre-ft/yr of recharge to Jakes Valley may move as interbasin flow to Railroad Valley.

Goshute and Steptoe Valleys comprise the Goshute Valley flow system. Recharge from precipitation to Goshute and Steptoe Valleys was estimated to be about 173,000 acre-ft/yr and ground-water evapotranspiration was estimated to be about 170,500 acre-ft/yr. About 2,000 acre-ft/yr were assumed to leave Goshute Valley as interbasin flow to the Great Salt Lake Desert flow system and about 500 acre-ft/yr as interbasin flow to Antelope Valley.

Spring, Antelope, and Tippett Valleys are part of the Great Salt Lake Desert flow system, which includes 13 more valleys in Utah. Total estimated recharge from precipitation to the three valleys in the study area was estimated at about 133,500 acre-ft/yr plus an estimated 500 acre-ft/yr as interbasin flow from Goshute Valley to Antelope Valley for a total recharge of 134,000 acre-ft/yr. Total ground-water evapotranspiration was estimated to be about 96,900 acre-ft/yr. The balance of 37,100 acre-ft/yr is proposed to move as interbasin flow into other valleys of the Great Salt Lake Desert system to the east.

Regional interbasin ground-water flow between the valleys and flow systems of eastern Nevada has been estimated by earlier studies, but the present study proposes interbasin flow in greater magnitude and, in some places, between valleys other than those previously suggested. The proposed areas of interbasin flow, or perhaps shift in hydrologic-basin boundaries that may more properly reflect the fact that hydrologic divides are not coincident with topographic divides, are based on geologic, geophysical, and hydrologic data. Regional interbasin ground-water flow from Long Valley to Jakes Valley and Railroad Valley has been suggested by previous investigations. Interbasin flow to Newark Valley from Long Valley has not been suggested by earlier studies, but is supported by analysis of the distribution of ground-water evapotranspiration in Newark Valley developed during the present study.

Interbasin flow from Butte Valley to Clover Valley had not been suggested by earlier studies, but is supported by thc large estimated recharge to Butte Valley, available water-level data, and geologic conditions in the area between Butte and Clover Valleys. Interbasin flow of about 4,000 acre-ft/yr from Spring Valley to Snake Valley was estimated by a previous study, but the volume was increased to an estimated 10,000 acre-ft/yr by the present study. Interbasin flow of about 5,000 acre-ft/yr from Antelope Valley and 7,000 acre-ft/yr from Tippett had been proposed by an earlier study; these amounts were increased by the present study and are supported by spring discharge measurements. New studies of ground-water budgets using the current methods for estimating ground-water recharge and evapotranspiration in the White River Valley and the rest of the Colorado flow system, as well as the other valleys of the Great Salt Lake Desert system, would allow the testing and substantiation of the new estimates of interbasin ground-water flow.

REFERENCES CITED

Billings, W.D., 1951, Vegetation zonation in the Great Basin of western North America: Union of Biological Science, Series B, p. 101-122.

Blaney, H.F., Ewing, P.A., Israelsen, O.W., Rohwer, C., and Scobey, F.C., 1938, Regional Planning, Part VI – The Rio Grande joint investigation in the upper Rio Grande Basin in Colorado, New Mexico, and Texas, 1936-1937, Part III – Water utilization: U.S. Bureau of Agricultural Engineering, National Resources Committee Report, v. 1, p. 293-427.

Blaney, H.F., Taylor, C.A., and Young, A.A., 1930, Rainfall penetration and consumptive use of water in Santa Ana River Valley and coastal plain: California Department of Public Works, Division of Water Resources Bulletin 33, 162 p.

Blaney, H.F., Taylor, C.A., Young, A.A., and Nickle, H.G., 1933, Water losses under natural conditions from wet areas in southern California: California Department of Public Works, Division of Water Resources Bulletin 44, Part I, 139 p.

Cardinalli, J.L., Roach, L.M., Rush, F.E., and Vasey, B.J., comps., 1968, State of Nevada hydrographic areas: Nevada Division of Water Resources map, scale 1:500,000.

Charlet, D.A., 1996, Atlas of Nevada conifers: Reno, University of Nevada Press, 320 p.

Coats, R.R., 1987, Geology of Elko County, Nevada: Nevada Bureau of Mines and Geology Bulletin 101, 112 p.

Czarnecki, J.B., 1997, Geohydrology and evapotranspiration at Franklin Lake playa, Inyo County, California *with a section on* Estimating evapotranspiration using the energy-budget eddy-correlation technique by David I. Stannard: U.S. Geological Survey Water-Supply Paper 2377, 75 p.

Daly, C., Neilson, R.P., and Phillips, D.L., 1994, A statistical-topographic model for mapping climatological precipitation over mountainous terrain: Journal of Applied Meteorology, v. 33, p. 140-158.

Dettinger, M.D., 1989, Distribution of carbonate-rock aquifers in southern Nevada and the potential for their development - Summary of findings, 1985-88: Carson City, Nev., Program for the Study and Testing of Carbonate-Rock Aquifers in Eastern and Southern Nevada, Summary Report No. 1, 37 p.

Dettinger, M.D., and Schaefer, D.H., 1996, Hydrogeology of extended terrains in the eastern Great Basin from geologic and geophysical models: U.S. Geological Survey Hydrologic Investigations Atlas HA-694-D, 1 sheet.

Dudley, W.W., 1967, Hydrogeology and ground-water flow system of the central Ruby Mountains, Nevada: University of Illinois, Urbana, unpublished Ph. D. dissertation, 106 p.

Duell, L.F.W., Jr., 1990, Estimates of evapotranspiration in alkaline scrub and meadow communities of Owens Valley, California, using the Bowen-ratio, eddy-correlation, and Penman-combination methods: U.S. Geological Survey Water-Supply Paper 2370-E, 39 p.

Eakin, T.E., 1950, Preliminary report on ground water in Fish Lake Valley, Nevada and California, *with a statement on* Reconnaissance land classification of Fish Lake Valley, Esmeralda County, Nevada by H.G. Mason: Nevada State Engineer, Water Resources Bulletin 11, 37 p.

———1960, Ground-water appraisal of Newark Valley, White Pine County, Nevada: Nevada Department of Conservation and Natural Resources, Ground-Water Resources - Reconnaissance Report 1, 33 p.

———1961, Ground-water appraisal of Long Valley, White Pine and Elko Counties, Nevada: Nevada Department of Conservation and Natural Resources, Ground-Water Resources-Reconnaissance Report 3, 35 p.

———1966, A regional interbasin groundwater system in the White River area, southeastern Nevada: Water Resources Research, v. 2, no. 2, p. 251-271.

Eakin, T.E., Hughes, J.L., and Moore, D.O., 1967, Water-resources appraisal of Steptoe Valley, White Pine and Elko Counties, Nevada: Nevada Department of Conservation and Natural Resources, Ground-Water Resources-Reconnaissance Report 42, 48 p.

Eakin, T.E., and Maxey, G.B., 1951a, Ground water in Ruby Valley, Elko and White Pine Counties, Nevada, *in* Eakin, T.E., Maxey, G.B., Robinson, T.W., Fredricks,

J.C., and Loeltz, O.J., 1951, Contributions to the hydrology of eastern Nevada: Nevada State Engineer, Water Resources Bulletin 12, p. 65-93.

———1951b, Ground water in Clover and Independence Valleys, Elko County, Nevada, *in* Eakin, T.E., Maxey, G.B., Robinson, T.W., Fredricks, J.C., and Loeltz, O.J., 1951, Contributions to the hydrology of eastern Nevada: Nevada State Engineer, Water Resources Bulletin 12, p. 99-125.

Eakin, T.E., Maxey, G.B., and Robinson, T.W., 1951a, Ground water in Goshute and Antelope Valleys, Elko County, Nevada, *in* Eakin, T.E., Maxey, G.B., Robinson, T.W., Fredricks, J.C., and Loeltz, O.J., 1951, Contributions to the hydrology of eastern Nevada: Nevada State Engineer, Water Resources Bulletin 12, p. 17-34.

Eakin, T.E., Maxey, G.B., Robinson, T.W., Fredricks, J.C., and Loeltz, O.J., 1951b, Contributions to the hydrology of eastern Nevada: Nevada State Engineer, Water Resources Bulletin 12, 171 p.

Gates, J.S., and Kruer, S.A., 1981, Hydrologic reconnaissance of the southern Great Salt Lake Desert and summary of the hydrology of west-central Utah: Utah Department of Natural Resources, Technical Publication 71, 55 p.

Gatewood, J.S., Robinson, T.W., Colby, B.R., Hem, J.D., and Halpenny, L.C., 1950, Use of water by bottom-land vegetation in lower Safford Valley, Arizona: U.S. Geological Survey Water-Supply Paper 1103, 210 p.

Glancy, P.A., 1968, Water-resources appraisal of Butte Valley, Elko and White Pine Counties, Nevada: Nevada Department of Conservation and Natural Resources, Ground-Water Resources-Reconnaissance Report 49, 50 p.

———1986, Geohydrology of the basalt and unconsolidated sedimentary aquifers in the Fallon area, Churchill County, Nevada: U.S. Geological Survey Water-Supply Paper 2263, 62 p.

Greenlee, D., 1992, Nevada annual data summary, water year 1992: Soil Conservation Service, Reno, Nev., 24 p.

Groeneveld, D.P., 1989, Shrub rooting and water acquisition on threatened shallow groundwater habitats in the Owens Valley, California: Symposium on Cheatgrass Invasion, Shrub Die-off, and Other Aspects of Shrub Biology and Management, Las Vegas, April 1989, p. 221-237.

Hardman, G., 1936, Precipitation map of Nevada: Nevada Agricultural Experiment Station, scale not given.

Harrill, J.R., 1971, Water-resources appraisal of the Pilot Creek Valley area, Elko and White Pine Counties, Nevada: Nevada Division of Water Resources, Reconnaissance Report 56, 46 p.

Harrill, J.R., Gates, J.S., and Thomas, J.M., 1988, Major ground-water flow systems in the Great Basin region of Nevada, Utah, and adjacent states: U.S. Geological Survey Hydrologic Investigations Atlas HA-694-C, 2 sheets.

Harrill, J.R., and Prudic, D.E., 1998, Aquifer systems in the Great Basin region of Nevada, Utah, and adjacent states—Summary report: U.S. Geological Survey Professional Paper 1409-A, 66 p.

Hose, R.K., and Blake, M.C., Jr., 1976, Geology and mineral resources of White Pine County, Nevada, Part I – Geology: Nevada Bureau of Mines and Geology Bulletin 85, 35 p.

Houghton, J.G., Sakamoto, C.M., and Gifford, R.O., 1975, Nevada's weather and climate: Nevada Bureau of Mines and Geology Special Publication 2, 78 p.

Kleinhampl, F.J., and Ziony, J.I., 1984, Geology of northern Nye County, Nevada: Nevada Bureau of Mines and Geology Bulletin 99A, 172 p.

LaPointe, D.D., Tingley, J.V., and Jones, R.B., 1991, Mineral resources of Elko County, Nevada: Nevada Bureau of Mines and Geology Bulletin 106, 226 p.

Lee, C.H., 1912, Water resources of a part of Owens Valley, California: U.S. Geological Survey Water-Supply Paper 294, 135 p.

Lerner, D.N., Issar, A.S., and Simmers, I., 1990, Groundwater recharge: International Association of Hydrogeologists, International Contributions to Hydrogeology, v. 8, 345 p.

Malek, E., Bingham, G.B., and McCurdy, G.D., 1990, Evapotranspiration from the margin and moist playa of a closed desert valley: Journal of Hydrology, v. 120, p. 15-34.

Maxey, G.B., and Eakin, T.E., 1949, Ground water in White River Valley, White Pine, Nye, and Lincoln Counties, Nevada: Nevada State Engineer, Water Resources Bulletin 8, 59 p.

———1951, Ground water in Railroad, Hot Creek, Reveille, Kawich, and Penoyer Valleys, Nye, Lincoln, and White Pine Counties, Nevada, *in* Eakin, T.E., Maxey, G.B., Robinson, T.W., Fredricks, J.C., and Loeltz, O.J., 1951, Contributions to the hydrology of eastern Nevada: Nevada State Engineer, Water Resources Bulletin 12, p. 129-171.

Meinzer, O.E., 1917, Geology and water resources of Big Smoky, Clayton, and Alkali Springs Valleys, Nevada: U.S. Geological Survey Water-Supply Paper 423, 167 p.

Mifflin, M.D., 1968, Delineation of ground-water flow systems in Nevada: University of Nevada, Desert Research Institute Technical Report H-W4, 89 p.

Mower, R.W., and Nace, R.L., 1957, Water consumption by water-loving plants in the Malad Valley, Oneida County, Idaho: U.S. Geological Survey Water-Supply Paper 1412, 33 p.

Mozingo, N.H., 1987, Shrubs of the Great Basin: Reno, University of Nevada Press, 342 p.

Nichols, W.D., 1992a, Energy budgets and resistance to energy transport in sparsely vegetated rangeland: Agricultural and Forest Meteorology, v. 60, p. 221-247.

Nichols, W.D., 1992b, The uncertainty of water budget estimates in the Great Basin, *in* Herrmann, R., ed., Managing water resources during global change: American Water Resources Association 28th Annual Conference and Symposium, Reno, Nev., Nov. 1-5, 1992, p. 309-317.

———1993, Estimating discharge of shallow groundwater by transpiration from greasewood in the northern Great Basin: Water Resources Research, v. 29, no. 8, p. 2771-2778.

———1994, Groundwater discharge by phreatophyte shrubs in the Great Basin as related to the depth to groundwater: Water Resources Research, v. 30, no. 12, p. 3265-3274.

Nichols, W.D., Laczniak, R.J., DeMeo, G.A., and Rapp, T.R., 1997, Estimated ground-water discharge by evapotranspiration, Ash Meadows National Wildlife Refuge, Nye County, Nevada, 1994: U.S. Geological Survey Water-Resources Investigations Report 97-4025, 13 p.

Plume, R.W., 1996, Hydrogeologic framework of the Great Basin region of Nevada, Utah, and adjacent states: U.S. Geological Survey Professional Paper 1409-B, 64 p.

Plume, R.W., and Carlton, S.M., 1988, Hydrogeology of the Great Basin region of Nevada, Utah, and adjacent states: U.S. Geological Survey Hydrologic Investigations Atlas HA694-A, scale 1:1,000,000.

Ponce, D.A., 1992, Bouguer gravity map of Nevada, Ely sheet: Nevada Bureau of Mines and Geology Map 99, 1 sheet.

Ponce, D.A., Morin, R.L., and Robbins, S.L., 1996, Bouguer gravity map of Nevada, Elko sheet: Nevada Bureau of Mines and Geology Map 107, 1 sheet.

Prudic, D.E., Harrill, J.R., and Burbey, T.J., 1995, Conceptual evaluation of regional ground-water flow in the carbonate-rock province of the Great Basin, Nevada, Utah, and adjacent states: U.S. Geological Survey Professional Paper 1409-D, 102 p.

Robertson, J.H., 1983, Greasewood (*Sarcobatus vermiculatus* (Hook.) Torr.): Phytologia, v. 54, no. 5, p. 309-324.

Robinson, T.W., 1958, Phreatophytes: U.S. Geological Survey Water-Supply Paper 1423, 84 p.

———1970, Evapotranspiration by woody phreatophytes in the Humboldt River valley near Winnemucca, Nevada: U.S. Geological Survey Professional Paper 491-D, 41 p.

Rush, F.E., 1968, Index of hydrographic areas: Nevada Division of Water Resources, Information Report 6, 38 p.

Rush, F.E., and Everett, D.E., 1966, Water-resources appraisal of Little Fish Lake, Hot Creek, and Little Smoky Valleys, Nevada: Nevada Department of Conservation and Natural Resources, Reconnaissance Report 38, 39 p.

Rush, F.E., and Kazmi, S.A.T., 1965, Water resources appraisal of Spring Valley, White Pine and Lincoln Counties: Nevada Department of Conservation and Natural Resources, Ground-Water Resources-Reconnaissance Report 33, 36 p.

Rush, F.E., Scott, B.R., Van Denburgh, A.S., and Vasey, B.J., comps., 1971, State of Nevada water resources and interbasin flows: Nevada Division of Water Resources map, scale 1:750,000.

Scott, B.R., 1971, Average annual evaporation: Nevada Division of Water Resources map, scale 1:2,000,000.

Shamberger, H.A., 1962, A proposed 10-year cooperative water resources program between the State of Nevada and the U.S. Geological Survey: Nevada Department of Conservation and Natural Resources, Water Resources - Information Report 4, 18 p.

Shreve, F., 1942, The desert vegetation of North America: Botanical Review, v. 7, no. 4, p. 195-246.

Shuttleworth, W.J., and Gurney, R.J., 1990, The theoretical relationship between foliage temperature and canopy resistance in sparse crops: Quarterly Journal of the Royal Meteorological Society, v. 111, p. 839-855.

Smith, J.L., Reece, B.D., and Medina, R.L., 2000, Data sets and related information for regional ground-water evapotranspiration in eastern Nevada: U.S. Geological Survey Open-File Report 99-242, CD-ROM.

Smith, R.M., 1976, Geology and mineral resources of White Pine County, Nevada. Part II, Mineral resources: Nevada Bureau of Mines and Geology Bulletin 85, p. 36-105.

Snyder, D.B., Healy, D.L., and Saltus, R.W., 1984, Bouguer gravity map of Nevada, Lund sheet: Nevada Bureau of Mines and Geology Map 80, 1 sheet.

Stannard, D.I., 1997, Estimating evapotranspiration using the energy-budget eddy-correlation technique, *in* Czarnecki, J.B., 1997, Geohydrology and evapotranspiration at Franklin Lake playa, Inyo County, California: U.S. Geological Survey Water-Supply Paper 2377, p. 25-36.

Thomas, J.M., Mason, J.L., and Crabtree, J.D., 1986, Ground-water levels in the Great Basin region of Nevada, Utah, and adjacent states: U.S. Geological Survey Hydrologic Investigations Atlas HA-694-B, 2 sheets.

Trimble, S., 1989, The sagebrush ocean: Reno, University of Nevada Press, 248 p.

Tschanz, C.M., and Pampeyan, E.H., 1970, Geology and mineral resources of Lincoln County, Nevada: Nevada Bureau of Mines and Geology Bulletin 73, 187 p.

Van Denburgh, A.S., and Rush, F.E., 1974, Water-resources appraisal of Railroad and Penoyer Valleys, east-central Nevada: Nevada Division of Water Resources, Reconnaissance Report 60, 61 p.

Watson, P., Sinclair, P., and Waggoner, R., 1976, Quantitative evaluation of a method for estimating recharge to the desert basins of Nevada: Journal of Hydrology, v. 31, p. 335-357.

Weeks, E.P., Weaver, H.L., Campbell, G.S, and Tanner, B.D., 1987, Water use by saltcedar and by replacement vegetation in the Pecos River floodplain between Acme and Artesia, New Mexico: U.S. Geological Survey Professional-Paper 491-G, 33 p.

White, W.N., 1932, A method of estimating ground-water supplies based on discharge by plants and evaporation from soil—Results of investigations in Escalante Valley, Utah, *in* Contributions to the hydrology of the United States 1932: U.S. Geological Survey Water-Supply Paper 659, p. 1-105.

Winograd, I.J., and Thordarson, William, 1975, Hydrogeologic and hydrochemical framework, south-central Great Basin, Nevada-California, with special reference to the Nevada Test Site: U.S. Geological Survey Professional Paper 712-C, 126 p.

Young, A.A., and Blaney, H.F., 1942, Use of water by native vegetation: California Department of Public Works, Division of Water Resources, Bulletin 50, 160 p.

Table C17. Estimated ground-water discharge by evapotranspiration from valleys in eastern Nevada study area, 1985 and 1989 [1]

	1985						1989					
			Evapotranspiration						Evapotranspiration			
	Average cover	Area (acres)	Annual average rate (feet)	Summer (acre-feet)	Winter (acre-feet)	Annual (acre-feet)	Average cover	Area (acres)	Annual average rate (feet)	Summer (acre-feet)	Winter (acre-feet)	Annual (acre-feet)
Antelope Valley												
Open water		0				0		0				0
Playa / Bare soil		279	0.150			(2)		19	0.150			(2)
<10 percent plant cover	0.046	10,894	.164	1,239	544	1,783	0.073	9,994	.479	3,367	1,422	4,789
10 to <20 percent plant cover	.126	108	1.200	93	36	129	.117	1,263	1.101	997	394	1,391
20 to <35 percent plant cover	.259	1	2.172	2	1	3	.242	4	2.082	6	2	8
35 to <50 percent plant cover							.398	1	2.489	1	0	1
Total		11,282		1,334	581	1,915		11,281		4,371	1,818	6,189
Butte Valley												
Open water		32				(3)		3				(3)
Playa / Bare soil		1,315	.150			197		213	.150			32
<10 percent plant cover	.059	64,990	.314	14,282	6,125	20,407	.081	58,335	.580	23,847	10,002	33,849
10 to <20 percent plant cover	.140	8,043	1.346	7,835	2,988	10,823	.129	15,050	1.221	13,241	5,131	18,372
20 to <35 percent plant cover	.248	1,977	2.110	3,107	1,062	4,169	.251	2,314	2.120	3,658	1,248	4,906
35 to <50 percent plant cover	.420	468	2.507	898	275	1,173	.419	623	2.507	1,195	367	1,562
≥50 percent plant cover	.595	176	2.585	355	101	456	.631	465	2.584	937	264	1,201
Total		77,001		26,477	10,551	37,225		77,003		42,878	17,012	59,922
Clover Valley												
Open water		7,347				(3)		3				(3)
Playa / Bare soil		2,072	.150			311		4,107	.150			616
<10 percent plant cover	.078	15,968	.566	6,381	2,658	9,039	.080	17,816	.601	7,572	3,142	10,714
10 to <20 percent plant cover	.138	22,551	1.328	21,661	8,279	29,940	.131	26,891	1.256	24,355	9,408	33,763
20 to <35 percent plant cover	.258	10,042	2.147	16,095	5,462	21,557	.270	6,930	2.189	11,354	3,818	15,172
35 to <50 percent plant cover	.422	4,009	2.509	7,700	2,359	10,059	.420	4,011	2.507	7,694	2,360	10,054
≥50 percent plant cover	.620	4,276	2.587	8,624	2,440	11,064	.686	6,508	2.584	13,166	3,649	16,815
Total		66,265		60,461	21,198	81,970		66,266		64,141	22,377	87,134
Goshute Valley												
Open water		8				(3)		1				(3)
Playa / Bare soil		4,197	.150			630		41	.150			6
<10 percent plant cover	.047	129,981	.192	17,384	7,578	24,962	.078	118,816	.537	44,894	18,902	63,796
10 to <20 percent plant cover	.134	1,178	1.275	1,084	417	1,501	.117	16,275	1.095	12,770	5,049	17,819
20 to <35 percent plant cover	.256	377	2.143	603	205	808	.264	491	2.168	796	269	1,065
35 to <50 percent plant cover	.414	95	2.502	181	56	237	.409	149	2.496	284	88	372
≥50 percent plant cover	.583	47	2.588	95	27	122	.688	109	2.579	220	61	281
Total		135,883		19,347	8,283	28,260		135,882		58,964	24,369	83,339

Table C17. Estimated ground-water discharge by evapotranspiration from valleys in eastern Nevada study area, 1985 and 1989 [1]—Continued

	1985						1989					
			Evapotranspiration						Evapotranspiration			
	Average cover	Area (acres)	Annual average rate (feet)	Summer (acre-feet)	Winter (acre-feet)	Annual (acre-feet)	Average cover	Area (acres)	Annual average rate (feet)	Summer (acre-feet)	Winter (acre-feet)	Annual (acre-feet)
Hot Creek Valley												
Open water		0				0		0				0
Playa / Bare soil		20	0.150			3		7	0.150			1
<10 percent plant cover	0.072	1,084	.490	374	157	531	0.077	1,167	.542	446	187	633
10 to <20 percent plant cover	.146	1,819	1.415	1,868	707	2,575	.146	1,857	1.406	1,894	717	2,611
20 to <35 percent plant cover	.253	620	2.127	984	335	1,319	.251	551	2.120	872	297	1,169
35 to <50 percent plant cover	.411	168	2.499	321	99	420	.414	123	2.501	235	72	307
≥50 percent plant cover	.613	67	2.586	134	38	172	.589	73	2.587	146	42	188
Total		3,778		3,681	1,336	5,020		3,778		3,593	1,315	4,909
Independence Valley												
Open water		211				(3)		1				(3)
Playa / Bare soil		15,162	.150			2,274		4,606	.150			691
<10 percent plant cover	.061	77,224	.355	19,225	8,167	27,392	.073	77,604	.507	27,712	11,614	39,326
10 to <20 percent plant cover	.131	9,494	1.250	8,565	3,307	11,872	.119	19,397	1.113	15,488	6,102	21,590
20 to <35 percent plant cover	.223	488	2.008	726	253	979	.250	892	2.118	1,408	481	1,889
35 to <50 percent plant cover	.382	4	2.467	8	2	10	.398	78	2.484	148	46	194
≥50 percent plant cover							.538	5	2.580	10	3	13
Total		102,583		28,524	11,729	42,527		102,583		44,766	18,246	63,703
Jakes Valley												
Open water		35				(3)		7				(3)
Playa / Bare soil		8	.150			(2)		4	.150			(2)
<10 percent plant cover	.080	16	.604	7	3	10	.088	81	.693	40	16	56
10 to <20 percent plant cover	.154	218	1.492	237	89	326	.140	285	1.352	279	107	386
20 to <35 percent plant cover	.248	122	2.112	193	66	259	.243	40	2.091	62	21	83
35 to <50 percent plant cover	.399	12	2.486	24	7	31						
≥50 percent plant cover	.638	5	2.586	10	3	13						
Total		416		471	168	639		417		381	145	526
Little Fish Lake Valley												
Open water		800				600		187				600
Playa / Bare soil		37	.150			6		31	.150			5
<10 percent plant cover	.078	1,803	.549	698	292	990	.078	2,890	.561	1,143	477	1,620
10 to <20 percent plant cover	.143	3,791	1.386	3,807	1,447	5,254	.141	3,845	1.356	3,776	1,439	5,215
20 to <35 percent plant cover	.260	1,178	2.156	1,897	642	2,539	.255	1,003	2.138	1,601	544	2,145
35 to <50 percent plant cover	.417	387	2.505	741	228	969	.402	187	2.489	355	110	465
≥50 percent plant cover	.587	176	2.586	353	101	454	.559	27	2.584	55	16	71
Total		8,172		7,496	2,710	10,812		8,170		6,930	2,586	10,121

Table C17. Estimated ground-water discharge by evapotranspiration from valleys in eastern Nevada study area, 1985 and 1989 [1]—Continued

		1985						1989					
				Evapotranspiration						Evapotranspiration			
		Average cover	Area (acres)	Annual average rate (feet)	Summer (acre-feet)	Winter (acre-feet)	Annual (acre-feet)	Average cover	Area (acres)	Annual average rate (feet)	Summer (acre-feet)	Winter (acre-feet)	Annual (acre-feet)
Little Smoky Valley													
	Open water		1				(3)		0				0
	Playa / Bare soil		600	0.150			90		45	0.150			7
	<10 percent plant cover	0.048	3,403	.232	553	236	789	0.062	3,227	.374	849	359	1,208
	10 to <20 percent plant cover	.145	953	1.396	965	365	1,330	.139	1,563	1.334	1,508	576	2,084
	20 to <35 percent plant cover	.268	865	2.185	1,414	476	1,890	.265	950	2.174	1,544	521	2,065
	35 to <50 percent plant cover	.410	298	2.497	569	175	744	.413	359	2.500	686	211	897
	≥50 percent plant cover	.687	192	2.582	388	108	496	.620	167	2.587	338	95	433
	Total		6,312		3,889	1,360	5,339		6,311		4,925	1,762	6,694
Long Valley													
	Open water		3				(3)		0				0
	Playa / Bare soil		985	.150			148		330	.150			49
	<10 percent plant cover	.066	19,253	.374	5,040	2,158	7,198	.077	19,934	.540	7,578	3,191	10,769
	10 to <20 percent plant cover	.119	1,633	1.113	1,304	514	1,818	.113	1,606	1.045	1,200	478	1,678
	20 to <35 percent plant cover	.218	8	1.983	11	4	15	.248	10	2.107	16	6	22
	35 to <50 percent plant cover							.397	2	2.486	3	1	4
	Total		21,882		6,355	2,676	9,179		21,882		8,797	3,676	12,522
Newark Valley													
	Open water		9,860				(3)		5				(3)
	Playa / Bare soil		13,920	.150			2,088		9,250	.150			1,388
	<10 percent plant cover	.068	32,742	.443	10,216	4,296	14,512	.062	45,826	.410	13,231	5,540	18,771
	10 to <20 percent plant cover	.136	18,970	1.301	17,832	6,842	24,674	.130	21,427	1.239	19,138	7,406	26,544
	20 to <35 percent plant cover	.256	5,476	2.142	8,756	2,974	11,730	.263	4,460	2.165	7,218	2,440	9,658
	35 to <50 percent plant cover	.414	1,513	2.501	2,895	891	3,786	.413	1,466	2.500	2,803	863	3,666
	≥50 percent plant cover	.638	784	2.584	1,580	445	2,025	.625	832	2.582	1,674	474	2,148
	Total		83,265		41,279	15,448	58,815		83,266		44,064	16,723	62,175
Railroad Valley (northern part)	Open water		16				50		31				50
	Playa / Bare soil		77,549	.150			11,632		47,822	.150			7,173
	<10 percent plant cover	.049	143,367	.232	23,243	9,959	33,202	.049	176,210	.224	27,535	11,883	39,418
	10 to <20 percent plant cover	.132	17,849	1.266	16,315	6,290	22,605	.135	15,339	1.290	14,293	5,493	19,786
	20 to <35 percent plant cover	.266	3,663	2.174	5,954	2,008	7,962	.262	3,910	2.163	6,320	2,137	8,457
	35 to <50 percent plant cover	.420	1,874	2.507	3,596	1,103	4,699	.418	1,801	2.505	3,453	1,060	4,513
	≥50 percent plant cover	.684	2,477	2.582	5,006	1,390	6,396	.636	1,682	2.587	3,396	956	4,352
	Total		246,795		54,114	20,750	86,546		246,795		54,997	21,529	83,749

Table C17. Estimated ground-water discharge by evapotranspiration from valleys in eastern Nevada study area, 1985 and 1989 [1]—Continued

	1985						1989					
			Evapotranspiration						Evapotranspiration			
	Average cover	Area (acres)	Annual average rate (feet)	Summer (acre-feet)	Winter (acre-feet)	Annual (acre-feet)	Average cover	Area (acres)	Annual average rate (feet)	Summer (acre-feet)	Winter (acre-feet)	Annual (acre-feet)
Ruby Valley												
Open water[4]		45,609				(5)		28				(5)
Playa / Bare soil[4]		4,688	0.150			703		4,367	0.150			655
<10 percent plant cover[4]	0.065	28,880	.406	8,258	3,481	11,739	0.069	68,313	.461	22,025	9,249	31,274
10 to <20 percent plant cover[4]	.146	32,930	1.410	33,687	12,736	46,423	.141	44,723	1.361	42,670	16,242	58,912
20 to <35 percent plant cover[4]	.256	20,611	2.141	32,932	11,188	44,120	.256	14,622	2.138	23,273	7,910	31,183
35 to <50 percent plant cover[4]	.420	7,380	2.508	14,163	4,343	18,506	.419	5,280	2.506	10,125	3,107	13,232
≥50 percent plant cover[4]	.608	5,046	2.586	10,161	2,887	13,048	.663	7,811	2.586	15,789	4,406	20,195
Ruby Lake/Marsh		14,637				22,000		14,637				22,000
Total		159,781		[4]99,201	[4]34,635	156,539		159,781		[4]113,882	[4]40,914	177,451
Spring Valley												
Open water		7,647				(3)		27				(3)
Playa / Bare soil		18,197	.150			2,730		38,147	.150			5,722
<10 percent plant cover	.054	102,598	.283	20,357	8,686	29,043	.048	103,189	.226	16,296	6,997	23,293
10 to <20 percent plant cover	.141	22,951	1.357	22,555	8,587	31,142	.141	14,541	1.359	14,315	5,449	19,764
20 to <35 percent plant cover	.260	10,166	2.154	16,358	5,541	21,899	.261	6,988	2.160	11,276	3,816	15,092
35 to <50 percent plant cover	.418	3,788	2.505	7,260	2,229	9,489	.419	2,750	2.506	5,273	1,618	6,891
≥50 percent plant cover	.632	2,889	2.585	5,824	1,643	7,467	.656	2,594	2.582	5,232	1,466	6,698
Total		168,236		72,354	26,686	101,770		168,236		52,392	19,346	77,460
Steptoe Valley												
Open water		7,328				(3)		166				(3)
Playa / Bare soil		12,402	.150			1,860		9,359	.150			1,404
<10 percent plant cover	.058	89,842	.322	20,289	8,636	28,925	.067	91,595	.418	26,916	11,377	38,293
10 to <20 percent plant cover	.145	28,384	1.396	28,744	10,892	39,636	.140	36,428	1.350	35,606	13,570	49,176
20 to <35 percent plant cover	.256	12,071	2.140	19,281	6,551	25,832	.259	14,044	2.152	22,576	7,652	30,228
35 to <50 percent plant cover	.418	3,984	2.505	7,635	2,345	9,980	.415	4,072	2.502	7,790	2,396	10,186
≥50 percent plant cover	.685	4,562	2.581	9,214	2,558	11,772	.654	2,909	2.583	5,870	1,645	7,515
Total		158,573		85,163	30,982	118,005		158,573		98,758	36,640	136,802
Tippett Valley												
Open water		0				0		0				0
Playa / Bare soil		1,090	.150			(2)		1	.150			(2)
<10 percent plant cover	.036	6,516	.102	463	203	666	.077	6,504	.529	2,423	1,019	3,442
10 to <20 percent plant cover	.124	155	1.174	131	51	182	.120	1,254	1.131	1,018	401	1,419
20 to <35 percent plant cover	.254	7	2.126	11	4	15	.249	9	2.115	14	5	19
35 to <50 percent plant cover							.374	1	2.460	1	0	1
Total		7,768		605	258	863		7,769		3,456	1,425	4,881

[1] Calculated values are not rounded, to minimize rounding errors in subsequent calculations.

[2] Deep water table; no ground-water evapotranspiration from playa/bare-soil areas.

[3] Supported by surface-water runoff; no ground-water component.

[4] Exclusive of Ruby Lake/Marsh.

[5] Ground-water component of open-water evaporation is included in Ruby Lake/Marsh category.

Table C18. Area and annual volume of precipitation in valleys of eastern Nevada study area [1]

Valley	Precipitation zone	Average precipitation (feet)	Area (acres)	Volume (acre-feet)
Antelope Valley	10	0.833	77,236	64,363
	11	.917	67,628	61,992
	12	1.000	63,017	63,017
	13	1.083	16,784	18,183
	14	1.167	15,696	18,312
	15	1.250	6,567	8,208
	16	1.333	4,014	5,352
	17	1.417	1,679	2,378
	18	1.500	1,714	2,570
	19	1.583	827	1,309
	20	1.667	520	867
Average precipitation		1.250		
Total			255,682	246,551
Butte Valley	10	.833	138,603	115,502
	11	.917	101,873	93,384
	12	1.000	181,753	181,753
	13	1.083	55,129	59,723
	14	1.167	48,432	56,504
	15	1.250	27,002	33,752
	16	1.333	24,541	32,721
	17	1.417	13,829	19,591
	18	1.500	10,548	15,822
	19	1.583	8,045	12,738
	20	1.667	11,966	19,943
	21	1.750	5,877	10,284
	22	1.833	6,375	11,687
	23	1.917	4,803	9,206
	24	2.000	7,912	15,825
	25	2.083	1,803	3,756
	26	2.167	1,319	2,858
	27	2.250	1,198	2,696
	28	2.333	1,354	3,159
Average precipitation		1.583		
Total			652,362	700,904
Clover Valley	12	1.000	135,088	135,088
	13	1.083	37,760	40,907
	14	1.167	45,071	52,583
	15	1.250	11,508	14,385
	16	1.333	8,762	11,682
	17	1.417	6,409	9,079
	18	1.500	4,345	6,518
	19	1.583	4,624	7,322
	20	1.667	7,613	12,688
	21	1.750	1,922	3,364

Table C18. Area and annual volume of precipitation in valleys of eastern Nevada study area [1]—Continued

Valley	Precipitation zone	Average precipitation (feet)	Area (acres)	Volume (acre-feet)
Clover Valley—Continued	22	1.833	2,925	5,362
	23	1.917	2,294	4,396
	24	2.000	3,213	6,426
	25	2.083	1,778	3,704
	26	2.167	1,353	2,932
	27	2.250	1,587	3,570
	28	2.333	1,598	3,728
	29	2.417	1,236	2,986
	30	2.500	1,885	4,713
	31	2.583	1,174	3,032
	32	2.667	2,233	5,955
	33	2.750	1,583	4,354
	34	2.833	1,925	5,454
	35	2.917	1,315	3,835
	36	3.000	711	2,132
	37	3.083	377	1,163
	38	3.167	555	1,757
	39	3.250	337	1,097
	40	3.333	935	3,117
Average precipitation		2.167		
Total			292,116	363,329
Goshute Valley	9	.750	30	22
	10	.833	263,098	219,247
	11	.917	93,149	85,387
	12	1.000	128,126	128,126
	13	1.083	32,499	35,207
	14	1.167	33,030	38,535
	15	1.250	19,352	24,190
	16	1.333	18,325	24,434
	17	1.417	8,334	11,807
	18	1.500	7,252	10,878
	19	1.583	3,390	5,368
	20	1.667	3,575	5,958
	21	1.750	605	1,060
	22	1.833	718	1,317
	23	1.917	365	700
	24	2.000	320	640
Average precipitation		1.375		
Total			612,168	592,876
Hot Creek Valley	6	.500	254,292	127,146
	7	.583	58,069	33,873
	8	.667	177,037	118,025
	9	.750	59,089	44,317
	10	.833	74,746	62,288

Table C18. Area and annual volume of precipitation in valleys of eastern Nevada study area [1]—Continued

Valley	Precipitation zone	Average precipitation (feet)	Area (acres)	Volume (acre-feet)
Hot Creek Valley—Continued	11	0.917	10,580	9,698
	12	1.000	12,118	12,118
	13	1.083	2,098	2,273
	14	1.167	1,997	2,329
	15	1.250	1,964	2,455
	16	1.333	1,545	2,060
	17	1.417	1,581	2,240
	18	1.500	1,953	2,929
	19	1.583	863	1,366
	20	1.667	570	950
Average precipitation		1.083		
Total			658,502	424,067
Independence Valley	10	.833	4,448	3,707
	11	.917	17,792	16,309
	12	1.000	202,825	202,825
	13	1.083	36,105	39,114
	14	1.167	35,579	41,509
	15	1.250	15,046	18,808
	16	1.333	16,401	21,868
	17	1.417	9,750	13,812
	18	1.500	10,082	15,123
	19	1.583	3,391	5,370
	20	1.667	6,493	10,821
	21	1.750	824	1,443
	22	1.833	615	1,127
	23	1.917	722	1,383
	24	2.000	598	1,196
Average precipitation		1.417		
Total			360,671	394,415
Jakes Valley	12	1.000	182,857	182,857
	13	1.083	26,996	29,245
	14	1.167	21,740	25,364
	15	1.250	9,822	12,278
	16	1.333	23,029	30,705
	17	1.417	2,676	3,790
	18	1.500	1,951	2,927
	19	1.583	821	1,299
	20	1.667	607	1,011
Average precipitation		1.333		
Total			270,499	289,476
Little Fish Lake Valley	8	.667	53,696	35,798
	9	.750	27,146	20,360
	10	.833	114,403	95,336
	11	.917	24,972	22,891
	12	1.000	32,626	32,626

Table C18. Area and annual volume of precipitation in valleys of eastern Nevada study area [1]—Continued

Valley	Precipitation zone	Average precipitation (feet)	Area (acres)	Volume (acre-feet)
Little Fish Lake Valley—Continued	13	1.083	7,553	8,182
	14	1.167	4,288	5,003
	15	1.250	3,419	4,273
	16	1.333	2,507	3,343
	17	1.417	2,275	3,222
	18	1.500	3,597	5,396
Average precipitation		1.083		
Total			276,482	236,430
Little Smoky Valley	6	.500	218,599	109,299
	7	.583	61,432	35,835
	8	.667	123,577	82,385
	9	.750	122,435	91,826
	10	.833	127,189	105,991
	11	.917	25,497	23,372
	12	1.000	26,125	26,125
	13	1.083	4,796	5,196
	14	1.167	4,279	4,992
	15	1.250	3,923	4,904
	16	1.333	5,461	7,281
	17	1.417	5,745	8,139
	18	1.500	6,088	9,133
	19	1.583	2,023	3,203
	20	1.667	3,407	5,678
Average precipitation		1.083		
Total			740,576	523,359
Long Valley	10	.833	66,702	55,585
	11	.917	49,757	45,611
	12	1.000	109,778	109,778
	13	1.083	31,746	34,392
	14	1.167	67,412	78,648
	15	1.250	37,729	47,161
	16	1.333	29,724	39,631
	17	1.417	9,700	13,742
	18	1.500	8,909	13,363
	19	1.583	2,783	4,406
	20	1.667	2,010	3,350
	21	1.750	1,149	2,010
	22	1.833	872	1,599
	23	1.917	658	1,261
	24	2.000	915	1,830
Average precipitation		1.417		
Total			419,844	452,367
Newark Valley	6	.500	1,796	898
	7	.583	5,872	3,425
	8	.667	75,423	50,282
	9	.750	35,042	26,281
	10	.833	53,543	44,619

Table C18. Area and annual volume of precipitation in valleys of eastern Nevada study area [1]—Continued

Valley	Precipitation zone	Average precipitation (feet)	Area (acres)	Volume (acre-feet)
Newark Valley—Continued	11	0.917	44,756	41,027
	12	1.000	139,558	139,558
	13	1.083	35,963	38,959
	14	1.167	29,337	34,227
	15	1.250	14,890	18,613
	16	1.333	14,000	18,667
	17	1.417	10,174	14,414
	18	1.500	9,918	14,877
	19	1.583	8,471	13,413
	20	1.667	8,972	14,953
	21	1.750	5,368	9,394
	22	1.833	4,551	8,344
	23	1.917	4,017	7,699
	24	2.000	4,776	9,551
	25	2.083	794	1,653
	26	2.167	759	1,644
	27	2.250	796	1,790
	28	2.333	506	1,182
Average precipitation		1.417		
Total			509,282	515,470
Railroad Valley (northern part)	6	.500	378,759	189,379
	7	.583	155,268	90,572
	8	.667	191,821	127,882
	9	.750	107,689	80,767
	10	.833	129,850	108,208
	11	.917	85,128	78,034
	12	1.000	71,651	71,651
	13	1.083	44,336	48,031
	14	1.167	47,187	55,052
	15	1.250	31,883	39,853
	16	1.333	36,215	48,286
	17	1.417	16,411	23,249
	18	1.500	14,036	21,053
	19	1.583	11,156	17,664
	20	1.667	19,538	32,563
	21	1.750	5,865	10,264
	22	1.833	4,995	9,157
	23	1.917	4,239	8,125
	24	2.000	4,718	9,435
	25	2.083	1,727	3,598
	26	2.167	1,093	2,368
	27	2.250	2,308	5,194
	28	2.333	3,798	8,863
Average precipitation		1.417		
Total			1,369,671	1,089,248

Table C18. Area and annual volume of precipitation in valleys of eastern Nevada study area [1]—Continued

Valley	Precipitation zone	Average precipitation (feet)	Area (acres)	Volume (acre-feet)
Ruby Valley	11	0.917	308	283
	12	1.000	161,087	161,087
	13	1.083	67,429	73,048
	14	1.167	123,654	144,263
	15	1.250	49,507	61,884
	16	1.333	50,472	67,296
	17	1.417	31,709	44,921
	18	1.500	27,042	40,563
	19	1.583	18,327	29,018
	20	1.667	16,631	27,719
	21	1.750	9,609	16,816
	22	1.833	7,561	13,861
	23	1.917	7,250	13,896
	24	2.000	7,047	14,093
	25	2.083	5,546	11,554
	26	2.167	5,201	11,270
	27	2.250	5,222	11,750
	28	2.333	6,922	16,152
	29	2.417	4,532	10,951
	30	2.500	4,352	10,880
	31	2.583	4,346	11,228
	32	2.667	6,722	17,926
	33	2.750	2,360	6,489
	34	2.833	2,115	5,992
	35	2.917	1,872	5,459
	36	3.000	2,754	8,263
	37	3.083	1,366	4,212
	38	3.167	1,583	5,014
	39	3.250	1,357	4,409
	40	3.333	4,433	14,776
	41	3.417	318	1,087
	42	3.500	168	588
	43	3.583	133	475
Average precipitation		2.250		
Total			638,935	867,223
Spring Valley	8	.667	106,811	71,208
	9	.750	72,758	54,568
	10	.833	248,999	207,498
	11	.917	107,803	98,820
	12	1.000	133,646	133,646
	13	1.083	68,709	74,434
	14	1.167	63,007	73,509
	15	1.250	46,419	58,024
	16	1.333	45,371	60,494
	17	1.417	29,512	41,809

Table C18. Area and annual volume of precipitation in valleys of eastern Nevada study area [1]—Continued

Valley	Precipitation zone	Average precipitation (feet)	Area (acres)	Volume (acre-feet)
Spring Valley—Continued	18	1.500	31,580	47,370
	19	1.583	16,305	25,817
	20	1.667	28,406	47,344
	21	1.750	6,818	11,932
	22	1.833	6,780	12,430
	23	1.917	6,280	12,037
	24	2.000	8,862	17,724
	25	2.083	5,073	10,569
	26	2.167	4,294	9,303
	27	2.250	4,912	11,052
	28	2.333	5,517	12,873
	29	2.417	4,795	11,588
	30	2.500	3,482	8,705
	31	2.583	3,588	9,269
	32	2.667	7,283	19,421
Average precipitation		1.667		
Total			1,067,010	1,141,444
Steptoe Valley	8	.667	134,239	89,493
	9	.750	111,476	83,607
	10	.833	228,925	190,770
	11	.917	76,121	69,778
	12	1.000	202,268	202,268
	13	1.083	82,523	89,400
	14	1.167	89,928	104,916
	15	1.250	50,044	62,555
	16	1.333	49,177	65,569
	17	1.417	41,810	59,231
	18	1.500	37,803	56,705
	19	1.583	21,133	33,461
	20	1.667	33,722	56,203
	21	1.750	11,579	20,264
	22	1.833	10,232	18,759
	23	1.917	8,298	15,905
	24	2.000	19,027	38,053
	25	2.083	6,937	14,452
	26	2.167	5,415	11,731
	27	2.250	5,459	12,282
	28	2.333	6,906	16,115
	29	2.417	2,238	5,407
	30	2.500	1,208	3,019
	31	2.583	1,813	4,683
	32	2.667	7,336	19,563
Average precipitation		1.667		
Total			1,245,617	1,344,189

Table C18. Area and annual volume of precipitation in valleys of eastern Nevada study area [1]—Continued

Valley	Precipitation zone	Average precipitation (feet)	Area (acres)	Volume (acre-feet)
Tippett Valley	9	0.750	1,992	1,494
	10	.833	102,228	85,190
	11	.917	43,695	40,054
	12	1.000	24,271	24,271
	13	1.083	13,382	14,497
	14	1.167	15,448	18,023
	15	1.250	6,627	8,284
	16	1.333	5,013	6,684
	17	1.417	3,398	4,814
	18	1.500	2,699	4,049
	19	1.583	1,863	2,950
	20	1.667	958	1,596
Average precipitation		1.208		
Total			221,574	211,906

[1] Calculated values are not rounded, to minimize rounding errors in subsequent calculations.

Table C19. Precipitation areas and volumes for selected precipitation zones and estimated recharge in valleys of eastern Nevada study area [1]

Precipitation range (inches)	Weighted average precipitation (inches)	Area in zone (acres)	Precipitation in zone (acre-feet per year)	Recharge factor	Estimated recharge (acre-feet per year)
Antelope Valley					
10-11	0.872	144,863	126,355	.008	1,011
12-15	1.055	102,064	107,720	.130	14,004
16-19	1.410	8,233	11,609	.144	1,672
20	1.667	520	867	.158	137
Total		255,680	246,551		16,824
Butte Valley					
10-11	.869	240,477	208,887	.008	1,671
12-15	1.062	312,316	331,732	.130	43,125
16-19	1.420	56,963	80,872	.144	11,646
20-28	1.864	42,607	79,414	.158	12,547
Total		652,363	700,905		68,989
Clover Valley					
12-15	1.059	229,427	242,962	.130	31,585
16-19	1.433	24,140	34,601	.144	4,983
20-33	2.075	32,393	67,210	.158	10,619
34-40	3.015	6,155	18,554	.626	11,615
Total		292,115	363,327		58,802
Goshute Valley					
9-11	.855	356,277	304,657	.008	2,437
12-15	1.061	213,006	226,057	.130	29,387
16-19	1.407	37,302	52,486	.144	7,558
20-24	1.733	5,584	9,675	.158	1,529
Total		612,169	592,875		40,911
Hot Creek Valley					
6-7	.515	312,361	161,019	.000	0
8-11	.729	321,452	234,328	.008	1,875
12-15	1.055	18,176	19,175	.130	2,493
16-19	1.447	5,942	8,595	.144	1,238
20	1.667	570	950	.158	150
Total		658,501	424,067		5,756
Independence Valley					
10-11	.900	22,239	20,016	.008	160
12-15	1.044	289,556	302,256	.130	39,293
16-19	1.418	39,624	56,173	.144	8,089
20-24	1.726	9,251	15,970	.158	2,523
Total		360,670	394,415		50,065
Jakes Valley					
12-15	1.034	241,415	249,744	.130	32,467
16-19	1.360	28,476	38,722	.144	5,576
20	1.667	607	1,011	.158	160
Total		270,498	289,477		38,203